D1600040

TRANSITION AND TURBULENCE

Publication No. 46
of the Mathematics Research Center
The University of Wisconsin–Madison

Academic Press Rapid Manuscript Reproduction

This work relates to Department of the Navy Research Grant N00014–91–G–0003 issued by the Office of Naval Research. The United States Government has a royalty-free license throughout the world in all copyrightable material contained herein.

TRANSITION AND TURBULENCE

Edited by

RICHARD E. MEYER

Mathematics Research Center
University of Wisconsin–Madison
Madison, Wisconsin

Proceedings of a Symposium
Conducted by The Mathematics Research Center
The University of Wisconsin–Madison
October 13–15, 1980

ACADEMIC PRESS 1981
A Subsidiary of Harcourt Brace Jovanovich, Publishers

NEW YORK LONDON TORONTO SYDNEY SAN FRANCISCO

ACADEMIC PRESS, INC.
111 Fifth Avenue, New York, New York 10003

United Kingdom Edition published by
ACADEMIC PRESS, INC. (LONDON) LTD.
24/28 Oval Road, London NW1 7DX

Library of Congress Cataloging in Publication Data
Main entry under title:

Transition and turbulence.

(Publication of the Mathematics Research Center, the University of Wisconsin--Madison ; no. 46)
Includes index.
1. Fluid dynamics--Congresses. 2. Transition flow--Congresses. 3. Turbulence--Congresses. I. Meyer, Richard E., Date. II. University of Wisconsin--Madison. Mathematics Research Center. III. Series.
QA3.U45 no. 46 [QA911] 510s [532'.05] 81-7903
ISBN 0-12-493240-1 AACR2

PRINTED IN THE UNITED STATES OF AMERICA

81 82 83 84 9 8 7 6 5 4 3 2 1

Contents

Senior Contributors

Numbers in parentheses indicate the pages on which the contributions begin.

T. Brooke Benjamin (25), Mathematical Institute, Oxford University, Oxford OX1 3LB, England

Fritz H. Busse (43), Department of Earth and Space Sciences, University of California at Los Angeles, Los Angeles, California 90024

Richard C. DiPrima (1), Mathematics Department, Rensselaer Polytechnic Institute, Troy, New York 12181

Michael Gaster (95), National Maritime Institute, Teddington TW1 OLW, England

John Laufer (63), Department of Aerospace Engineering, University of Southern California, Los Angeles, California 90007

Joseph T. C. Liu (167), Division of Engineering, Brown University, Providence, Rhode Island 02912

John L. Lumley (215), Sibley School of Mechanical and Aerospace Engineering, Cornell University, Ithaca, New York 14853

Michio Nishioka (113), University of Osaka Prefecture, Osaka, Japan

Stephen A. Orszag (127), Mathematics Department, Massachusetts Institute of Technology, Cambridge, Massachusetts 02139

Eli Reshotko (147), Department of Mechanical and Aerospace Engineering, Case Western Reserve University, Cleveland, Ohio 44106

Philip G. Saffman (149), Firestone Laboratory, California Institute of Technology, Pasadena, California 91125

J. Trevor Stuart (77), Department of Mathematics, Imperial College, London, SW7 2BZ, England

Preface

This volume collects invited lectures presented at a Symposium on Transition and Turbulence in fluids in Madison, Wisconsin, on October 13–15, 1980 under the auspices of the Mathematics Research Center of the University of Wisconsin–Madison, sponsored by the United States Army under Contract No DAAG29–80–C–0041 and supported by the Office of Naval Research, U. S. Navy, Grant N00014–81–G–0003, and the National Science Foundation under Grant MCS–8006171.

The symposium was devoted to a review of the insights gained over the past two decades into the relation between transition and turbulence in fluids and on the importance of this relation for the understanding of many real fluid motions. I am greatly indebted to the contributors for the articles here putting on record a research conference which deeply impressed many of those attending. Invited lectures were also given by Professors Gary Brown, Hans Liepmann, and Hassan Nagib, whose articles could not be available in time for inclusion here. Hans Liepmann's lecture was a memorable experience for those who had the fortune to be present, and reference may be made to *American Scientist,* **Vol. *67*,** pp. 221–228, 1969, for some of the topics he discussed. Some of those discussed by Hassan Nagib, including a striking demonstration of how turbulence can be suppressed promptly by robbing it of the third dimension, are reported in a paper "Interaction of free-stream turbulence with screens and grids: the balance between turbulence scales" by N. M. Nagib and J. Tan-atichat now in press in the *Journal of Fluid Mechanics*.

I owe a special debt to John Laufer, Hans Liepmann, and Bill Pritchard for the good advice which proved to be the key to the success of the conference. Thanks are due also to Gladys Moran for the faultless handling of its details and to Judith Siesen for putting this volume together and compiling its index.

Richard E. Meyer

Transition in Flow between Rotating Concentric Cylinders

R. C. DiPrima

1. INTRODUCTION

In this paper we will discuss recent experimental and theoretical work on the instabilities and transitions of a viscous incompressible fluid between concentric rotating cylinders as the speed of the inner cylinder is increased. We will assume that the cylinders are of infinite length. Thus, in the references to theoretical work we neglect end effects which can change bifurcations into imperfect bifurcations. Recent work on end effects (with references) is discussed briefly in [1] and in the paper by T. Brooke Benjamin [2] in this volume.

Let R_1, R_2 and Ω_1, Ω_2 denote the radii and angular velocities of the inner and outer cylinders, respectively. Also, we let $d = R_2-R_1$ be the gap between the cylinders and, for discussion of the experiments, we let L denote the height of the fluid in the cylinders. We will only consider the case for which the outer cylinder is at rest so $\Omega_2 = 0$. Dimensionless parameters that are of special interest are

radius ratio	$\eta = R_1/R_2$,
aspect ratio	L/d ,
Reynolds number	$R = \Omega_1 R_1 d/\nu$,
Taylor number	$T = 2R^2(1-\eta)/(1+\eta)$.

ISBN 0-12-493240-1

In the discussion that follows we will assume that all quantities have been made dimensionless using the scales d for length, d^2/ν for time except that in reporting frequencies the scale Ω_1 is used, and $R_1\Omega_1$ for velocity.

The discussion is somewhat complicated by the fact that the theory is for $L/d = \infty$ while all experiments are done for finite values of L/d. Thus, Couette flow, which has only an azimuthal velocity component, is an exact solution of the Navier Stokes equations for all R for $L/d = \infty$; however, it is not a solution of these equations for L/d finite. Nevertheless, experimenters speak of Couette flow with the understanding that for moderate values of L/d end effects are small and for most practical purposes (such as a torque calculation) Couette flow is an adequate approximation. We will adopt such an attitude.

For an experimental apparatus with η near 1 (say $\eta > 0.8$) and for a moderate value of L/d (say $L/d \geq 20$), it is generally recognized that there are at least the following transitions as R is increased.

1. There is an $R_c(\eta, L/d)$ at which Couette flow becomes unstable. The instability leads to a steady axisymmetric flow which has a periodic cellular structure in the axial direction. The wavelength in the axial direction is nearly $2d$, i.e., the cells are nearly square. This flow is usually referred to as Taylor-vortex flow in honor of G. I. Taylor [3] for his classical paper in 1923 in which he calculated R_c for several values of η, confirmed his calculations by experiments, and observed the cellular secondary flow. Cole [4] has shown experimentally for a range of radius ratios $\eta = 0.894$-0.954 that R_c is very insensitive to L/d for $L/d > 8$. Of course R_c depends strongly on η. For example, according to the theoretical calculations, R_c is 185 for $\eta = 0.95$ and is 94.7 for $\eta = 0.8$.

The Taylor-vortex flow was calculated by Davey [5] in 1962 for several values of η and for $R-R_c$ small. Davey followed the method of amplitude expansions developed by Stuart [6] and Watson [7]. His results for the torque associated with the Taylor-vortex flow are in good agreement with experimental measurements.

2. There is an $R_w(\eta, L/d)$ at which Taylor-vortex flow becomes unstable. This instability leads to a wavy-vortex flow which has a definite azimuthal wave number and moves with a definite wave velocity in the azimuthal direction. The boundaries between adjacent cells are wavy, thus, the name wavy-vortex flow. Such flows were described by Taylor [3]. They were first photographed in 1956 by Schultz-Grunow and Hein [8]. Transition to wavy-vortex flow has been discussed in detail in a fundamental paper by Coles [9].

Cole [4] has shown that R_w is sensitive to L/d. For η near 1 theoretical predictions of R_w are only realistic for $L/d > 40$ and differences (from theory) of 15% can be expected for $L/d < 20$. Also, R_w is sensitive to η. For example, Cole finds that $R_w/R_c = 1.24$ for $\eta = 0.954$ and $L/d = 20$ with the ratio decreasing for larger values of L/d. In contrast, Snyder [10] and Snyder and Lambert [11] find that for $\eta = 1/2$ and $L/d \cong 30$ the Taylor-vortex regime exists and is stable for values of R as high as $10R_c$.

The first theoretical work on the instability of Taylor-vortex flow was that of Davey, DiPrima, and Stuart [12] in 1968. They derived a set of nonlinear equations for the time dependent amplitudes of axisymmetric and non-axisymmetric disturbances to the basic Couette flow. They found for the case of a small gap ($\eta \to 1$) that there was an R_w slightly higher than R_c at which the Taylor-vortex flow becomes unstable and the instability indeed leads to a wavy-vortex motion. More complete (and confirming) calculations have been carried out by Eagles [13]. We will return to this work in Section 3.

The wavy-vortex regime is more rich than has been implied by the preceding two paragraphs. Among the several observations made by Coles [9] in an apparatus with $\eta = 0.874$ and $L/d = 27.9$, there are two that we want to mention. First, it is possible to have wavy-vortex flows with different numbers of cells and waves at the same value of R - the final state depends upon how it is reached. This is also true in the axisymmetric regime, but there the number of states (cells) is far fewer. See DiPrima and Eagles [14] and the references of that paper for a discussion of the axisymmetric problem. Second, there are many transitions as the Reynolds number is varied. Let $(R/R_c, n, k)$ denote a wavy-vortex flow with n cells and k waves at Reynolds number R. Then as R is increased slowly Coles found the following sequence of transitions: (1,28,0), (1.25,28,4), (2.17,24,5), (3.21,22,5), (3.57,22,6), (6.66,22,5), and (8.50,22,4). In addition to these transitions, Coles observed other transitions as the speed of the inner cylinder was raised and lowered. We will not try to discuss the question of state selection and of transitions within the wavy-vortex regime.

3. There is an $R_q(\eta, L/d)$ at which the periodic motion (in time) becomes quasiperiodic as evidenced by the power spectrum of the velocity at a point having two incommensurate frequencies.

4. There is an $R_t(\eta, L/d)$ at which the flow field can be described as turbulent in the sense that the power spectrum of the velocity at a point has no spectral peaks. Even in this regime $(R > R_t)$ the flow field still maintains a periodic structure in the axial direction as was first noted by Pai [15].

Velocity power spectrum measurements which showed the existence of R_q and R_t were first carried out by Swinney and his co-workers as summarized in [16]. This and more recent experimental work will be discussed in Section 2. Theoretical work on the transitions to wavy-vortex flow and the quasiperiodic wavy-vortex flow and the calculation of such flows will be discussed in Section 3. A brief summary is given in Section 4.

2. EXPERIMENTAL OBSERVATIONS

We start by summarizing the experimental observations of Swinney and his co-workers as given in Fenstermacher, Swinney, and Gollub [16]. The experimental apparatus had a geometry $\eta = 0.877$ and $L/d = 20$. The upper fluid surface was free and the lower surface was fixed. Laser-Doppler techniques were used to measure the radial component of velocity at a point as a function of time for R in the range $5.4 < R/R_c < 45$. The wavy-vortex state was one with 17 cells in the axial direction and 4 waves in the azimuthal direction. For $\eta = 0.877$, the theoretical ($L/d = \infty$) value of R_c is 119.1. A summary of the transitions as R is increased is given in Table 2 of [16]. An adaptation of that table is given in Table 1.

Table 1. Transitions in Flow Between Concentric Cylinders for $\eta = 0.877$, $L/d = 20$ as reported in [16]

R/R_c	Remark
< 1	Couette flow
1 - 1.2	Steady Taylor-vortex flow
1.2 - 10.1	Wavy-vortex flow with frequency $f_1 \cong 1.3$
10.1 - 19.3	Quasiperiodic wavy-vortex flow with frequencies $f_1 \cong 1.3$ and $f_2 \cong 0.9$
$\cong$ 12.	Broad band (B) appears in the spectrum at $f \cong 0.45$. Flow is weakly turbulent with sharp spectral components
19.3	Frequency f_2 disappears. Flow is weakly turbulent with sharp spectral components.
21.9	Frequency f_1 disappears. Flow is weakly turbulent.

First, we want to mention that the value of R_c corresponds to a dimensionless wavenumber $a = 3.13$, whereas for the experiments the axial wavelength at the mid-height of the cylinder is 2.36d corresponding to a value of $a = 2.66$.

Second, the value of R_w was determined by visual observation, not by laser-Doppler techniques, and is only an approximate value. However, it is consistent with values given by Cole [4] for similar values of η and L/d. Third, it is interesting to note that once the 17/4 state was selected, it was possible to maintain it for $5.4 < R/R_c < 22$. This is somewhat in contrast to the transitions observed by Coles [9] in the wavy-vortex regime.

For these experiments, $R_w \cong 1.2R_c$, $R_q \cong 10R_c$, and $R_t \cong 22R_c$. The first frequency f_1 is relatively constant with increasing R, decreasing slightly for $5 < R/R_c < 15$ and then increasing slightly for $15 < R/R_c < 22$. The second frequency f_2 is a monotone slightly increasing function of R. The frequencies f_1 and f_2 appear to be incommensurate. For $R/R_c > 21.9$ the flow is weakly turbulent but continues to have a periodic structure in the axial direction as can be seen from photographs (g) and (h) of Figure 6 in [16].

We mentioned in the Introduction that the axial wave number for Taylor-vortex flow and the axial and azimuthal wave numbers for wavy-vortex flow are not unique, but rather depend on the way in which the final state is reached. This is also true for the second frequency, f_2. The relationship between f_1 and f_2 has been described by Gorman and Swinney [17]. They find that (1) the f_2 mode modulates the wavy-vortex mode both in amplitude and frequency, (2) the magnitude of f_2 differs with the number of azimuthal waves, (3) there are multiple values of f_2 for a wavy-vortex flow with the same number of waves, and (4) different values of f_2 give different spatial configurations of the modulation. For wavy-vortex flows with 4 waves the values of f_1 and f_2 at the onset of f_2 are given in Table 2. The data is taken from Table 1 of [17]. The 17/4 state is the one summarized in Table 1.

Table 2. Values of f_1 and f_2 at onset of f_2 mode. $\eta = 0.88$, $L/d = 20$. States with an even (odd) number of cells correspond to a rigid (free) upper surface.

cells/waves	R/R_c	f_1	f_2
16/4	11.5	1.32	0.44
17/4	10.2	1.38	0.82
18/4	9.6	1.30	1.16

Walden and Donnelly [18] have carried out a series of experiments in a cylinder apparatus with $\eta = 0.876$ for values of L/d in the range $18 < L/d < 80$ in order to assess dependence of flow states on L/d. The fluid was free at the upper surface. Visual observations were made using a glass outer cylinder and electrical measurements were made using a pair of gold-plated brass cylinders with ion collectors on the outer cylinder.

Their observations can be summarized in part as follows: (1) R_c is not affected by L/d unless there are only a few cells; (2) R_w is sensitive to L/d and increases with decreasing L/d; (3) R_t at which there is a transition from periodic or quasiperiodic to aperiodic (weakly turbulent flow) is sensitive to L/d, $R_t/R_c \cong 22$ for $L/d \cong 20$ and $R_t/R_c \cong 26$ for $L/d = 80$; and (4) for $L/d > 25$ and for $R > 28R_c$ a sharply defined spectral peak reappears in the power spectrum at about $f = 1.4$ and persists for values of R as large as $36R_c$. While observations (1) and (2) are not new and observation (3) is probably not surprising, observation (4) does indicate a new phenomenon, yet to be understood. We note that these observations are consistent with those of Fenstermacher et al [16] who did not find a reemergent spectral peak for $L/d = 20$.

In another paper, Donnelly, Park, Shaw, and Walden [19] have reported several transitions at values of R just slightly greater than R_c that have not previously been reported experimentally nor mentioned in any theoretical work. For these experiments $\eta = 0.876$, $L/d \cong 80$, and the upper surface is free. The flow was observed by use of an ion apparatus and visually (with measurement by lamp and photocell reflectance) by use of polymeric flakes in water.

The wavy-vortex flow appears at $R/R_c \cong 1.2\ R_c$ and has two waves (k=2) in the azimuthal direction. Surprisingly, at $R \cong 1.35R_c$ the power spectrum contains many broadened peaks at multiples of the frequency associated with a wavy-vortex flow with one wave (k=1). This state corresponds to dislocations in the cell boundaries; for example, a spiral cell between two wavy cells. The "dislocation activity" persists until $R \cong 2.1R_c$, then the power spectrum corresponds to a wavy-vortex flow with six waves (k=6). This periodic flow persists until $R \cong 3.3R_c$ when the power spectrum again shows a dislocation activity. With increasing R there are alternate ranges of R for which the flow is periodic or perhaps quasiperiodic and in states of dislocation activity. Hysteresis effects also are present. Dislocation activity is not observed for $L/d < 40$.

The experimental observations that we have just discussed reveal a variety of phenomena including transitions from steady flow to steady flow, from steady flow to periodic flow, from periodic flow to quasiperiodic flow, from periodic or quasiperiodic flow to weak turbulence, from a periodic flow to a time-dependent nonperiodic flow and the reverse transition, and so on. It is possible to construct systems of several nonlinear ordinary differential equations which, as a parameter varies, can exhibit one or more such transitions. But what is required is a mathematical model that can exhibit such transitions and truly represent the flow field. Some results are available, but much remains to be done. In the next section we discuss the theoretical results.

3. MATHEMATICAL THEORY

Almost all of theoretical work is limited to the prediction of R_c, the calculation of the Taylor-vortex flow, the prediction of R_w, and the calculation of the wavy-vortex flow. However, there has been some work on transitions in the wavy-vortex regime and at least one attempt to simulate the quasiperiodic flow for $R > R_q$ and the transition to weak turbulence for $R > R_t$.

Since R_w is fairly close to R_c for $\eta \to 1$, it is reasonable to hope that for this geometry perturbation methods for $R-R_c$ small can be used for the calculation of the Taylor-vortex flow, R_w, and the calculation of the wavy-vortex flow for some range of $R > R_c$. It is unlikely perturbation techniques can be used, at least not without a great deal of work, to predict the flow field in the quasiperiodic regime and the transition to weak turbulence.

For purpose of discussion, we consider only the azimuthal component of velocity $u_\theta(r,\theta,z,t)$, where r,θ, and z are the usual cylindrical coordinates and t is the time. If we assume the flow is periodic in the axial direction with wave number a and also periodic in the azimuthal direction with wave number k, we can write

$$u_\theta(r,\theta,z,t) = V(r) + \sum_{m=-\infty}^{\infty} \sum_{n=0}^{\infty} \sum_{j=1}^{\infty} [A_{cmnj}(t)v_{cmnj}(r)\cos naz + A_{smnj}(t)v_{smnj}(r)\sin naz]\, e^{imk\theta} \quad . \qquad (1)$$

Here $V(r)$ is the Couette velocity and the v's are the eigenfunctions of the linear stability problems corresponding to modes proportional to $\exp(imk\theta)\cos naz$ and $\exp(imk\theta)\sin naz$.

In the following, we shall always take a to be the axial wave number corresponding to R_c. Then for the linearized equations we know

$$A_{c011}(t) \text{ and } A_{s011}(t) \sim e^{\sigma t} \quad , \qquad (2)$$

where σ is real and $\sigma > 0$ for $R > R_c$. Also, we know for given k

$$A_{c111}(t) \text{ and } A_{s111}(t) \sim e^{\nu t} \quad , \qquad (3)$$

where ν is complex and $\nu_r > 0$ for $R > R_{kc} > R_c$. Values of R_{kc} are given in Table 3 for two values of η. Note that $\eta = 0.877$ is the value of η for the experiments summarized in Table 1. Also note that the critical values of R, R_{kc}, for non-axisymmetric modes ($k \neq 0$) are very close to R_c which corresponds to an axisymmetric mode. This is both good and bad. Good in the sense that the effect of the non-axisymmetric modes can be important for $R-R_c$ small, and bad in the sense that for $R-R_c$ positive and small many modes may grow according to linear theory.

Table 3. Values of R_{kc}/R_c for $\eta = 0.9512$ and $\eta = 0.877$

k	$\eta = 0.9512$, $a = 3.127$, $R_c = 187.3$	$\eta = 0.877$, $a = 3.13$, $R_c = 119.1$
1	1.002	1.007
2	1.008	1.028
3	1.020	1.068
4	1.036	1.130
5	1.057	1.228
6	1.085	1.391
7	1.122	1.712
8	1.170	3.032

In the analysis of Davey, DiPrima and Stuart [12] the simplest possible model that could allow Taylor-vortex flows and wavy-vortex flows was constructed using the fundamental modes represented by A_{c011}, A_{s011}, A_{c111}, A_{s111}, $A_{c-111} = \tilde{A}_{c111}$, and $A_{s-111} = \tilde{A}_{s111}$, where a tilda is used to denote the complex conjugate. First harmonics of these modes and the mean are generated by quadratic interactions of the fundamental modes; in turn, the fundamental modes interact with the mean and the first harmonics to generate the second harmonics and a correction to the radial dependence of the fundamental modes at cubic order in the amplitudes, and so on. At cubic order, the amplitudes satisfy the following set of nonlinear equations with $A_c = A_{c011}$, $A_s = A_{s011}$, $B_c = A_{c111}$, and $B_s = A_{s111}$:

$$\begin{aligned} dA_c/d\tau &= (\sigma + a_1 A_c^2 + a_1 A_s^2 + a_3 |B_c|^2 + a_4 |B_s|^2) A_c \\ &\quad + (a_5 B_c \tilde{B}_s + \tilde{a}_5 \tilde{B}_c B_s) A_s \ , \\ dB_s/d\tau &= (\nu + b_1 |B_s|^2 + b_2 |B_c|^2 + b_3 A_s^2 + b_4 A_c^2) B_s \\ &\quad + (b_3 - b_4) B_c A_s A_c + (b_1 - b_2) \tilde{B}_s B_c^2 \ , \end{aligned} \tag{4}$$

with similar equations for A_s and B_c. Also, the equations for $\tilde{B}_c$ and $\tilde{B}_s$ are the complex conjugates of those for B_c and B_s, respectively. The parameters σ (real) and ν (complex) are the growth rates of axisymmetric and non-axisymmetric disturbances, respectively, as noted in equations (2) and (3). The a's and b's are definite numbers that are calculated as part of the analysis.

The amplitude equations admit the possibility of a number of flow fields depending upon the values of the a's and b's: Couette flow, Taylor vortices, pure non-axisymmetric mode, wavy vortices, and spiral vortices. Of special interest is the possible transition from Couette flow to Taylor-vortex flow to wavy-vortex flow as R is increased. For this sequence we need only the A_c and B_s equations, namely equations (4) with $A_s = 0$ and $B_c = 0$. These equations have the possible solutions: Couette flow $A_c = B_s = 0$, Taylor-vortex flow $A_c \neq 0$, $B_s = 0$, pure non-axisymmetric mode $A_c = 0$, $B_s \neq 0$, and wavy-vortex flow $A_c \neq 0$, $B_s \neq 0$. Calculations by Davey et al [12] for the special case $\eta \to 1$ confirm the desired sequence - when Taylor-vortex flow becomes unstable wavy-vortex flow exists and is stable. Eagles [13] has extended the analysis to fifth order in the amplitudes and has calculated the necessary coefficients for $\eta = 0.9512$ without the approximation $\eta \to 1$. We summarize in Table 4 the results of his calculations for R_w.

Table 4. Values of T_{kc} and T_w for $\eta = 0.9512$, $a = 3.127$. ($T_c = 1754$, $R_c = 187.3$) $T_w(3)$ and $T_w(5)$ correspond to cubic and quintic amplitude equations.

k	1	2	3	4
T_{kc}	1761	1783	1823	1881
R_{kc}/R_c	1.002	1.008	1.020	1.036
$T_w(3)$	1924	1928		1945
$R_w(3)/R_c$	1.047	1.048		1.053
$T_w(5)$	1946	1954		1981
$R_w(5)/R_c$	1.053	1.055		1.063

Thus, if only perturbations of Couette flow are allowed that are axisymmetric with wave number $a = 3.127$ and non-axisymmetric with wave numbers $a = 3.127$ and, say, $k = 4$ we have the following situation: (i) Couette flow becomes unstable at $R_c = 187.3$ with the instability leading to a Taylor-vortex flow, (ii) in the absence of the axisymmetric mode, Couette flow would become unstable to the non-axisymmetric mode at $R = 1.036\ R_c$; however, the subsequent pure non-axisymmetric mode is unstable in the presence of the axisymmetric mode, (iii) the Taylor-vortex flow becomes unstable at $R_w = 1.06\ R_c$ with the instability leading to a wavy-vortex flow, (iv) the wavy-vortex flow exists and is stable for all $R > R_c$. There are several observations and limitations about the model that should be mentioned.

1. The wavy-vortex solution corresponds to a periodic mixed state $A_c \neq 0$, $B_s \neq 0$ of the cubic amplitude equations (4) with $A_s = B_c = 0$. Amplitude equations of this form can have mixed state solutions which can be unstable with the instability leading to a solution with a second time frequency and hence a quasiperiodic solution; see, for example, Langford [20]. However, the values of the a's and the b's as calculated by Eagles [13] for the true problem do not allow this possibility. Thus, this model, even if extended beyond its likely range of validity to $10R_c$, cannot yield a quasiperiodic flow as observed in the experiments.

2. For the cubic amplitude equations and for $k = 4$, a small change of 2 or 3% in the values of the a's and the b's calculated by Eagles [13] could cause the wavy-vortex solution to disappear for sufficiently large $R > R_w$. If so, then for these values of R the only stable solution would be the pure non-axisymmetric mode ($B_s \neq 0$, $A_c = 0$). It would be interesting to know how the fifth order terms in the amplitude equations affect the possible disappearance of the wavy-vortex solution. However, it should be noted that pure non-axisymmetric modes do not appear to be observed in practice.

3. Implicit in the derivation of the amplitude equations (4) is the assumption that σ and ν_r are both small. From a practical point, they are small since R_{kc} is very near R_c; however they are not simultaneously zero. In this sense, the theory is not rational; but it does appear possible to embed the analysis in a larger class of problems with a rational theory - work in progress by R. C. DiPrima and J. Sijbrand.

4. The analysis allows only axisymmetric modes (a,0) and non-axisymmetric modes (a,k). However, it is clear from Table 4 that the Taylor-vortex flow is nearly simultaneously unstable to non-axisymmetric modes with $k = 1, 2, 3, \cdots$, respectively. How does a wavy vortex with a definite azimuthal wave number k emerge? Moreover, for a given (a,k) the harmonics (a,mk) will have a "life of their own" according to linear theory and should have independent amplitudes rather than being considered as forced modes. To deal with these questions requires the inclusion of many fundamental modes - for a given R all the modes that can grow according to linear theory. For $\eta = 0.9512$, consideration of the linear stability problem for the modes (na,mk) with $a = 3.127$, $k = 1$, $n = 1, 2, 3$, and $m = 0, 1, \cdots, 8$ shows that all these modes † (102 in number) will

---

† Note that (na,0) provides two modes $\cos naz$ and $\sin naz$, while (na,mk) provides four modes $\exp(i \pm imk\theta)\cos naz$ and $\exp(\pm imk\theta)\sin naz$. If only the axisymmetric modes and non-axisymmetric modes proportional to $\cos naz$ and $\sin naz$, respectively, are considered, then the number is cut in half.

have lives of their own by $R = 3R_c$. The situation is only slightly better for $\eta = 0.877$, but with still decreasing η the number of modes that must be considered decreases rather rapidly.

The torque calculations of Eagles [21] for $\eta = 0.9512$ for wavy-vortex flows with one, two, three, and four azimuthal waves are somewhat relevant to these questions. Eagles found that the wavy vortex with four waves gives values of the torque in very good agreement with experimental observations. These calculations confirm the apparent preference of the dynamical system for a wavy-vortex flow with four waves, but we note that wavy-vortex flows with different number of waves have been observed.

Richtmyer [22] has extended the analysis to include ten and then fourteen modes associated with the wave numbers (a,0), (a,1), (a,2), and (a,3), as compared to the six modes associated with the wave numbers (a,0) and (a,k) considered by Davey et al [12]. For a ten-dimensional manifold and quintic amplitude equations, his results for $\eta = 0.9512$ confirm those of Davey et al [12] and Eagles [13], but they also add an interesting new piece of information. He finds

1. Taylor vortices appear at $T = 1753.1$ and are stable for $T < 1971.5 = 1.06R_c$.

2. Wavy vortices with $k = 1$ appear at $T = 1971.5$ and are stable for $T < 1985 = 1.064R_c$.

3. Wavy vortices with $k = 2$ appear at $T = 1985$ and are always stable for the range of values of T considered, namely $T < 3000 = 1.308R_c$.

This is the first calculation showing a transition from one wavy-vortex flow to another wavy-vortex flow. It is tempting to conjecture that if more modes could be included, one would find perhaps several transitions from one wavy vortex to another as R is increased over a small interval until a preferred state, possibly one with four waves, is reached. Also, stable helical modes were found (these were also found by Davey et al [12]). It is not clear that these helical modes can be reached as R is slowly increased, but they may be relevant to the dislocation activity with spiral cells observed by Donnelly et al [19].

For $\eta = 0.6$ and for the cubic amplitude equations, fourteen modes were used and the analysis was carried to $T \cong 446T_c$ or $R \cong 21R_c$ with $T_c \cong 2572$. For this value of η, the mode corresponding to four waves (k=4) is stable according to linear theory for the range of T considered and hence need not be included. The calculations show that the Taylor-vortex loses stability to a wavy-vortex flow with three waves somewhere between $11.2R_c$ and $15.8R_c$. This wavy-vortex flow is stable to $R \cong 21R_c$. No stable modes or periodic solutions were found for $R > 21R_c$ which suggests that the instability is a subcritical bifurcation.

Herbert [23] has developed a fully computational approach to calculate Taylor-vortex flows by expanding in powers of the amplitude of the vortex flow and then to study the stability of the Taylor-vortex flow to non-axisymmetric modes. His calculations to date are for the small gap case, $\eta \to 1$. He has calculated the Taylor-vortex flow to 15th order in the amplitude, and the stability analysis is carried out to fifth order in the amplitude. He finds that the Taylor-vortex flow is unstable to a k=1 non-axisymmetric mode, and out of phase by $\pi/2$ in the axial direction compared to the Taylor-vortex flow, at $R = 1.048R_c$. The value given by Davey et al [12] for $\eta \to 1$ and the k=1 mode is $R = 1.036R_c$. Herbert also found that Taylor-vortex flow was unstable to wavy-vortex modes with $k = 2, 3, \cdots$ at an increasing, but closely grouped, sequence of values of R. This is in agreement with our earlier discussion.

At the meeting of the Division of Fluid Dynamics of the American Physical Society (Cornell University, November, 1980), Herbert reported on work in progress on the stability of wavy-vortex flows to wavy vortices of a different azimuthal wave number. Again, the calculations are for $\eta \to 1$ with $a = 3.127$. He found the wavy-vortex flow with one wave that comes in at $R = 1.048R_c$ becomes unstable at $R = 1.053R_c$ with the instability leading to a wavy-vortex flow with three waves (k = 3). This wavy vortex persists against perturbation by other wavy-vortex modes to the highest value of R for which calculations were carried, namely $R = 1.086R_c$. This result is not in disagreement with that of Richtmyer for $\eta = 0.9512$, who found a transition from the $k = 1$ wavy vortex to the $k = 2$ wavy vortex since his analysis did not allow the possibility of a $k = 1$ to $k = 3$ transition. Herbert also found other transitions as the Reynolds number was decreased and then increased over the interval $R_c < R < 1.086R_c$. This is the first analytical work to develop a transitions table similar (but of course much more limited) to the transition table given by Coles [9].

C. A. Jones [24] has calculated R_w for several values of η by calculating the Taylor-vortex flow using the Galerkin method and then studying the stability of the Taylor-vortex flow to wavy-vortex disturbances. Thus, the approach is similar to that of Herbert and different from that used by Davey et al [12], Eagles [13], and Richtmyer [22] where the basic flow is Couette flow. The approach taken by Jones and by Herbert avoids the difficulty for η near one of the many unstable modes for Couette flow when R is only slightly greater than R_c.

Jones finds that (1) for $\eta > 0.8$, R_w corresponds to a non-axisymmetric mode with one wave in the azimuthal direction (k=1) which is consistent with the results of other investigators; (2) R_w increases with decreasing η, for example $R_w/R_c \cong 1.06$ for $\eta = 0.95$ and $R_w/R_c \cong 1.20$ for $\eta = 0.80$; (3) for η decreasing from $\eta = 0.8$ to 0.75, R_w increases very rapidly and corresponds to a k=2 mode at $\eta \cong 0.78$ and a k=3 mode at $\eta \cong 0.76$; (4) for $\eta < 0.75$ approximately, no instability of the Taylor-vortex flow was found for $T < 12{,}500$, and (5) for $\eta > 0.8$ and $R > R_w$ wavy-vortex modes with wave numbers greater than one have larger growth rates, for example, for $\eta = 0.8756$ and at $R = 1.5R_c$ the fastest growing wavy-vortex mode has six azimuthal waves. The latter observation may be relevant in a competition between modes which leads to a wavy-vortex flow with usually four, five, or six waves when R is moderately greater than R_c. The results (3) and (4) are consistent with the calculations of Richtmyer [22].

In regard to the rapid increase of R_w with decreasing η, K. Park and R. J. Donnelly reported on work in progress at the meeting of the Division of Fluid Dynamics of the American Physical Society (Cornell University, November, 1980) that is of interest. Experimental observations in an apparatus with $\eta = 0.5$ and $L/d \cong 25$ show no instability of the Taylor-vortex flow for values of T up to 1,340,000 ($R \cong 21R_c$).

Yahata [25, 26] has used an expansion of the form (1) to simulate the velocity field. However, rather than use the eigenfunctions v_{cmnj} and v_{smnj} of the linear stability problem, he uses sets of complete functions satisfying the necessary boundary conditions. The truncation of the series (1) is quite severe with only the Fourier components (a,0) proportional to sin az, (a,4) proportional to cos az, and (0,0) retained. The axial component of velocity and the pressure are solved for in terms of the radial and azimuthal components of velocity and then six expansion functions in the radial variable are used for the (a,0) and (a,4) modes with eight being used for the (0,0) mode. The resulting 32 amplitude equations are integrated numerically for the case $\eta = 0.875$ and $a = 2.50$.

The power spectrum of the fundamental component of the radial velocity for the (a,4) mode shows the following: for $R/R_c = 7.97$ there is a sharp frequency component at $f_1 \cong 1.5$ which presumably corresponds to the frequency of about 1.3 observed by Fenstermacher et al [16], but there is also frequency component at $f \cong 2$ which is not present in the experiments; for $R/R_c = 15.94$ there are frequency components at $f_1 \cong 1.5$ and $f_2 \cong 1.1$ which approximate frequencies $f_1 \cong 1.3$ and $f_2 \cong 0.9$ observed by Fenstermacher et al; for $R/R_c = 22.31$ the f_2 component disappears but the f_1 component remains; and for $R/R_c = 23.91$ there is a broad component at f_1. The behavior of the model is surprisingly similar to the experimental observations of Fenstermacher et al. Nevertheless, it is difficult to believe that the actual flow field can be correctly calculated by such a severely truncated series for even moderate values of $R > R_c$ yet alone for $R = 20R_c$.

In a later paper, Yahata [27] has used eight expansion functions in the radial variable for each of the Fourier components which yields a set of 40 amplitude equations. In this case, he finds a third frequency at $R/R_c = 20.72$, and hence has transitions from periodic motion to quasiperiodic motion with two fundamental frequencies to quasiperiodic motion with three fundamental frequencies to chaotic motion.

Finally, we wish to call attention to a recent report by D. Rand [28]. The paper, which is somewhat more abstract than other papers referred to in this section, provides a theoretical analysis of the quasiperiodic (doubly periodic) regime observed by Fenstermacher et al [16]. Possible modulation patterns for the f_2 mode modulation of the primary f_1 mode are derived from group theoretical ideas. Experiments by Gorman and Swinney [29] confirm the predicted modulation patterns.

4. SUMMARY

The use of laser-Doppler techniques, as exemplified in the work of Swinney and his co-workers [16], have led to significant advances in our understanding of the transition to turbulence for the flow between rotating cylinders. There can be little doubt about a sequence of transitions as R increases that involves, at least, steady flow → periodic flow → quasiperiodic flow → chaotic flow. The relevance of these observations within the framework of the theory of dynamical systems and in contrast to early conjectures that transition to turbulence will occur as an infinite sequence (or at least many) of hydrodynamical instabilities is discussed in Section 6 of [16].

Also, as noted earlier, and as shown by Cole [4] and Donnelly and his co-workers [18, 19], most transitions and phenomena such as remergent spectral peaks depend strongly on the parameters η and L/d. To complete the picture of this dependence requires additional experimental work.

In this paper it has not been possible to discuss all of the recent relevant experimental work for transition in the flow between rotating cylinders, so we would like to at least call attention, without discussion, to work by Barcilon, Brindley, Lessen, and Mobbs [30] and Koschmieder [31].

The range of Reynolds numbers required to show the rich variety of phenomena, say $R_c < R < 22R_c$ for $\eta = 0.876$, makes it difficult to hope that any attack on the governing nonlinear partial differential equations which requires a small parameter (soft nonlinear effects) can be completely successful. Even so, there are still open questions such as state selection which can be considered (at least in part) by using a small parameter $(R-R_c)$. The analytical-numerical work of Herbert [23] and Jones [24] appears to have considerable potential for dealing with the Taylor-vortex and wavy-vortex regimes. However, it is not clear how formidable it is to extend the numerics to $10R_c$ - $20R_c$. It is possible that the invariant manifold work of Richtmyer [22] could provide important information for a suitably chosen value of η such that the number of linearly unstable modes is sufficiently few, but yet sufficient to allow important interactions. Some value of η in the interval $0.6 < \eta < 0.877$ may be appropriate. However, it should be noted that his computational procedure with a fourteen-dimensional manifold was beginning to reach the capacity of the CRAY computer. It is possible, and desirable, to extend the calculations of Yahata [25-27] to include more Fourier components. However, the number of nonlinear coefficients goes up very rapidly as more modes are allowed and the data handling and numerical problems rapidly become difficult and the cost of the computation becomes expensive.

Nevertheless, it is fair to say that considerable progress has been made in the last few years in developing the theory for the flow between rotating concentric cylinders. The use of analytical-numerical techniques should yield even more satisfactory results in the next few years. However, remember that we are dealing with a classic system of nonlinear partial differential equations for a four-dimensional velocity - pressue field as a function of three-space variables and time, and it must be expected that the analysis and numerical work required to extract the desired information will not be easy. Finally, we remind the reader that the complicating feature of end effects has not been considered in theoretical work discussed in the review.

REFERENCES

1. DiPrima, R. C. and H. L. Swinney, Instabilities and transition in flow between concentric rotating cylinders, in Hydrodynamic Instabilities and the Transition to Turbulence (H. L. Swinney and J. P. Gollub, eds.), Springer, Berlin, Heidelberg, New York, 1981.
2. Benjamin, T. B., New observations in the Taylor experiment, Proceedings of Symposium on Transition and Turbulence, Mathematics Research Center, University of Wisconsin-Madison, October, 1980.
3. Taylor, G. I., Stability of a viscous liquid contained between two rotating cylinders, Philos. Trans. R. Soc. London A 223, 289-343 (1923).
4. Cole, J. A., Taylor-vortex instability and annulus-length effects, J. Fluid Mech., 75, 1-15 (1976).

5. Davey, A., The growth of Taylor vortices in flow between rotating cylinders, J. Fluid Mech., 14, 336-368 (1962).
6. Stuart, J. T., On the non-linear mechanics of wave disturbances in stable and unstable parallel flows, Part 1, The basic behaviour in plane Poiseuille flow, J. Fluid Mech., 9, 353-370 (1960).
7. Watson, J., On the non-linear mechanics of wave disturbances in stable and unstable parallel flows, Part 2, The development of a solution for plane Poiseuille flow and for plane Couette flow, J. Fluid Mech., 9, 371-389 (1960).
8. Schultz-Grunow, F. and H. Hein, Beitrag zur Couetteströmung, Z. Flugwiss, 4, 28-30 (1956).
9. Coles, D., Transition in circular Couette flow, J. Fluid Mech., 21, 385-425 (1965).
10. Snyder, H. A., Wavenumber selection at finite amplitude in rotating Couette flow, J. Fluid Mech., 35, 273-298 (1969).
11. Snyder, H. A. and R. B. Lambert, Harmonic generation in Taylor vortices between rotating cylinders, J. Fluid Mech., 26, 545-562 (1966).
12. Davey, A., R. C. DiPrima, and J. T. Stuart, On the instability of Taylor vortices, J. Fluid Mech., 31, 17-52 (1968).
13. Eagles, P. M., On stability of Taylor vortices by fifth-order amplitude expansions, J. Fluid Mech., 49, 529-550 (1971).
14. DiPrima, R. C. and P. M. Eagles, Amplification rates and torques for Taylor-vortex flows between rotating cylinders, Phys. Fluids, 20, 171-175 (1977).
15. Pai, S. I., Turbulent flow between rotating cylinders, National Advisory Committee for Aeronautics Technical Note No. 892 (1943).
16. Fenstermacher, P. R., H. L. Swinney, and J. P. Gollub, Dynamical instabilities and the transition to chaotic Taylor vortex flow, J. Fluid Mech., 94, 103-129 (1979).

17. Gorman, M. A. and H. L. Swinney, Visual observation of the second characteristic mode in a quasiperiodic flow, Phys. Rev. Lett., 43, 1871-1875 (1979).
18. Walden, R. W. and R. J. Donnelly, Reemergent order of chaotic circular Couette flow, Phys. Rev. Lett., 42, 301-304 (1979).
19. Donnelly, R. J., K. Park, R. Shaw, and R. W. Walden, Early nonperiodic transitions in Couette flow, Phys. Rev. Lett., 44, 987-989 (1980).
20. Langford, W. F., Interactions of Hopf and pitchfork bifurcations, in Bifurcation Problems and Their Numerical Solution, Birkhäuser, Basel, Boston, 1980, 103-134.
21. Eagles, P. M., On the torque of wavy vortices, J. Fluid Mech., 62, 1-9 (1974).
22. Richtmyer, R. D., Invariant manifolds and bifurcations in the Taylor problem, Private communication.
23. Herbert, T., Numerical studies on nonlinear hydrodynamic stability by computer-extended perturbation series, Proceedings of the 7th International Conference on Numerical Methods in Fluid Dynamics, Stanford, June 23-27, 1980, (to appear).
24. Jones, C. S., Nonlinear Taylor vortices and their stability, J. Fluid Mech., 92, (1981).
25. Yahata, H., Temporal development of the Taylor vortices in a rotating fluid. Prog. Theor. Phys. Suppl., 64, 176-185 (1978).
26. Yahata, H., Temporal development of the Taylor vortices in a rotating fluid II., Prog. Theor. Phys., 61, 791-800 (1979).
27. Yahata, H., Temporal development of the Taylor vortices in a rotating fluid. III, To be published.
28. Rand, D., The pre-turbulent transitions and flows of a viscous fluid between concentric rotating cylinders, Report of the Mathematics Institute, University of Warwick, Coventry, England, June 1980.

29. Gorman, M. and H. L. Swinney, Spatial characteristics of modulated waves in circular Couette flow - experiment and theory, Bulletin of the Am. Phys. Soc., 25, 1089 (1980).

30. Barcilon, A., J. Brindley, M. Lessen, and F. R. Mobbs, Marginal instability in Taylor-Couette flows at very high Taylor number, J. Fluid Mech., 94, 453-463 (1979).

31. Koschmieder, E. L., Transition from laminar to turbulent Taylor vortex flow, in Laminar-Turbulent Transition (R. Eppler and H. Fasel, eds.) Springer, Berlin, Heidelberg, New York, 1980.

Acknowledgment: This work was partially supported by the Army Research Office and by the Fluid Mechanics Branch of the Office of Naval Research. The author would like to express his thanks to Mr. Jonathan Sandberg for his assistance with some of the numerical calculations.

Department of Mathematical Sciences
Rensselaer Polytechnic Institute
Troy, New York 12181

New Observations in the Taylor Experiment

T. Brooke Benjamin

INTRODUCTION

Over half a century of inquiry has been devoted to the phenomena observed in G. I. Taylor's experiment on Couette flow between concentric circular cylinders. The phenomena have long been a cornerstone of concepts about hydrodynamic stability and the origins of turbulence, and interest in them has been greatly reinvigorated during the last decade in consequence of new practical and theoretical findings. The experiment is nevertheless a continuing source of perplexity, like Pandora's box, unfolding successive complications and attendant difficulties of interpretation while still exciting hopes for a final understanding of the richly various phenomena exhibited.

In most rehearsals of the Taylor experiment, including those whose outcome will be summarized below, the inner cylinder rotates at a continuously adjustable, constant angular speed Ω and the outer cylinder is stationary. A suitable Reynolds number, the sole dynamical parameter for the fluid motions, is then given by $R = \Omega r_1 d/\nu$, where r_1 is the radius of the inner cylinder, $d = r_2 - r_1$ the width of the fluid-filled annulus, and ν the kinematic viscosity. The geometric parameters affecting the flow are $\eta = r_1/r_2$ and the aspect ratio $\Gamma = \ell/d$, where ℓ is the length of the annulus. In many previous experiments, the fluid has been a liquid bounded above by a free surface, and this feature

ISBN 0-12-493240-1

undoubtedly had some bearing on what was observed although its precise influence is difficult to estimate. In the present experiments on the other hand, the end conditions are virtually symmetric: the bottom of the annular space is a fixed wall, and the top is the (stationary) surface of a PTFE collar whose distance ℓ from the bottom is continuously adjustable.

As is well known, the motion observed in any version of the Taylor apparatus is more or less featureless at small R, having no detectible radial or longitudinal component except perhaps close to the ends of the annulus. But as R is raised into a narrow, quasi-critical range, typically somewhat less than 100, an array of steady toroidal vortices (Taylor cells) is developed. The quasi-critical range of R is predicted quite closely by the idealized theory originally propounded by Taylor, which takes the annulus to be infinitely long and the motion to be an arbitrarily periodic perturbation from the uniform, circular Couette flow that is possible in this model. A strictly periodic array of cells is thus represented as a bifurcation from, and exchange of stability with, the basic Couette flow. It is now generally accepted, however, that in any real experiment the incipience of cellular motion is a process spreading from the ends of the annulus and depending continuously on R. Striking experimental demonstrations of this fact are now available (e.g. Kusnetsov et al. [1] Fig. 7). The suggestion was made [2], [3] that the process of cell development in a long annulus can be viewed as a simple 'softening' of the supercritical, symmetric bifurcation described by the idealized theory (i.e. as the unfolding of a symmetric bifurcation according to the canonical model of catastrophe theory, determined by a single perturbation parameter); but this rudimentary interpretation is unsatisfactory, giving no insight into the mechanism whereby the array of cells observed experimentally is determined as a function of η and, above all, of Γ. In fact the solution set for the hydrodynamic problem at large Γ has a prolific multiplicity in the quasi-critical range of R, and a rationale for the cell-selection process is consequently a matter of considerable intricacy [4].

In all cases, when R is raised to a higher threshold value, say R_w, which unlike the first is in principle sharply defined respective to each experimental value of η and Γ, the flow becomes unsteady with its cellular structure perturbed by circumferential travelling waves. It is already well recognized that R_w is sensitive to the values of η and Γ, generally increasing as they decrease, but the precise nature of this dependency, particularly on Γ, has yet to be fully explored. The implications of previous observations are obscured by lack of information about the end conditions imposed and about the particular array of Taylor cells whose stability limit is estimated.

Upon further increases of R the temporal fluctuations become progressively more complicated, including periodic components with several incommensurate frequencies and a growing residuum of noise, but the spatial organization of the flow in the form of Taylor cells may still be largely preserved. Eventually, at sufficiently high values of R, a plainly turbulent state of motion is observed, but it may still exhibit vestiges of cellular structure and regular pulsations at certain frequencies. Much recent interest has focussed on these developments at high R in the Taylor experiment, but consensus about their exact nature has yet to be established (e.g. the finding in refs. 5, 6 and 7 are different in important respects).

I shall outline as follows the results of three recent series of experiments on aspects of the Taylor experiment. My collaborator T. Mullin deserves principal credit for this experimental work, which is reported in full elsewhere [8], [9], [10]. The apparatus that was mainly used by us has $\eta = 0.615$, a value smaller than in most previous experiments (i.e. this is far from the 'narrow-gap' case often subsumed for convenience in the idealized theory), and Γ can be varied continuously over a range of comparatively small values. Perhaps due largely to the somewhat unusual proportions of this apparatus, several remarkable, hitherto unreported effects were made prominent. But we believe that corresponding effects are always present to some significant degree in the Taylor experiment, being thus intrinsic to a fully rational interpretation of the observed phenomena, even if these effects may be difficult to measure precisely in an apparatus with greater η and Γ.

MULTIPLE STEADY FLOWS

Previous experiments have established that for values of R some way above the quasi-critical range, numerous different stable steady flows are realizable in the Taylor experiment. For example, photographs have been published showing five different cellular flows realized as stable states at the same R and with the same dimensions of the annulus [11]. According to a rigorous qualitative theory that has been developed in several papers to account for such observations, [8], [12], [13], [14] the multiplicity of the complete solution set for the hydrodynamic problem in this case was at least 9, since necessarily there were 4 unstable, and therefore unobservable, flows complementing the 5 stable flows observed. (In general, the existence of N stable time-independent solutions implies the existence of $N - 1$ others which represent unstable flows.)

For any η and Γ, the steady hydrodynamic problem is well known to have a unique solution, which is stable, if R is sufficiently small. The flow that is evolved by gradual increases in R beyond the quasi-critical range has been termed the primary mode, and when subject to symmetric end conditions it is found always to develop an even number N of Taylor cells at high R, except when Γ is excessively small and a single-cell structure may be developed [8]. Primary modes with $N = 2,4,6,\ldots$, which number depends on Γ if η is fixed, are such that the axisymmetric spiralling motions in the cells have the normal sense of rotation, radially inwards near both end walls.

On the other hand, the term secondary modes has been introduced for steady flows that are stable when R exceeds some critical value but do not survive gradual reductions in r to smaller values. The selection process determining N as a function of Γ for the primary mode involves interactions and reversals of roles with secondary modes that also have an evey number of normally spiralling cells. This process occurs over narrow ranges of Γ, and includes hysteresis phenomena (with respect to variations in R) which are small but, with care, quite detectable. Measurements illuminating the change-over of the primary mode from a two-cell to a four-cell form with increasing Γ have already been reported [11], and studies of the four-six and subsequent changes are proceeding.

Stable steady flows featuring an odd number of cells in the symmetric apparatus, therefore having two possible realizations with an abnormally spiralling cell at one end or the other, have been called anomalous modes. This term also includes flows with an even number of cells but with abnormal spiralling in both end-cells. Except for the one-cell mode which has other special properties [8], these flows apparently have no parametric connection with the primary mode, thus remaining true secondary modes over the whole range of Γ wherein they are realizable.

Figure 1 reproduces from reference 8 a set of measurements of the stability limits for anomalous modes with 2 to 7 cells. The critical speed at which the respective mode collapses catastrophically is plotted as a function of Γ, having been estimated as the sequel to very gradual decreases in speed from much higher values. (The critical speed is here expressed by $R_o = \Omega_o r_1^2/\nu$, to be multiplied by $d/r_1 = 0.626$ to recover the preceding definition of R.) The experimental curves are reminiscent of a result according to the idealized theory, namely the critial R, for bifurcation from perfect Couette flow, expressed as a function of axial wavelength which may here be compared to ℓ/N with $N = 2,\dots,7$. The resemblance is deceptive, however because the minima of the curves are at values of R_c

over twice those indicated by the theory. Thus ends effects appear to be enormous in separating the anomalous modes from any counterparts of them that might be identified in the idealized theoretical model.

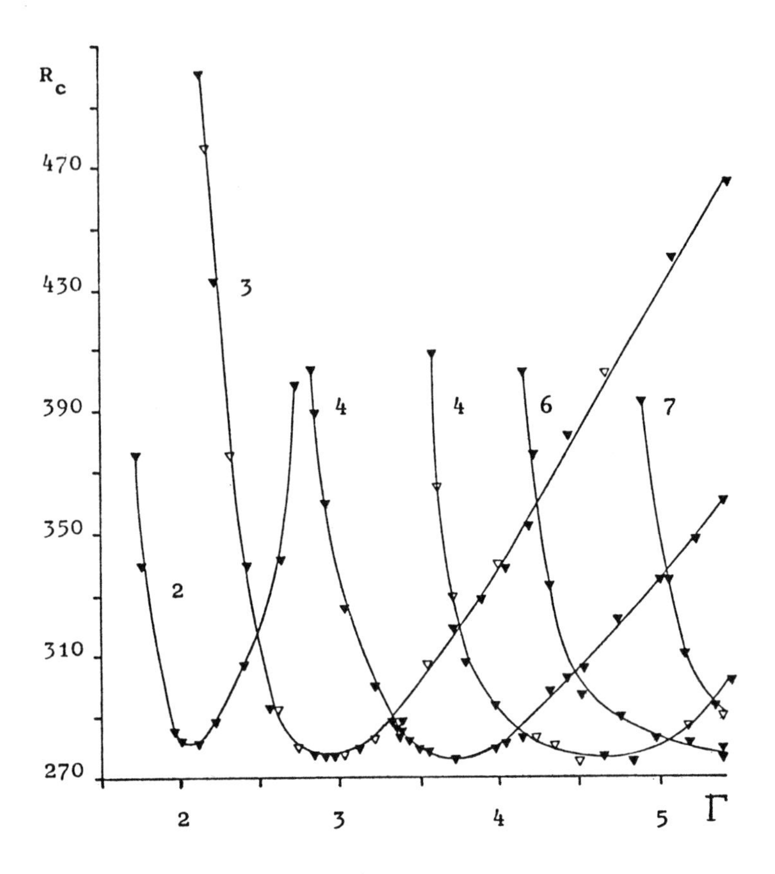

Figure 1. Lower stability limit for anomalous modes with 2 - 7 Taylor cells.

TRANSITION TO OSCILLATORY MOTION

In the second programme of experiments [9], careful measurements have been made on the high-speed limits of stability for various cellular flows, and again some surprising behaviour has come to light. A chosen cellular mode having been established as a steady state in the apparatus set at a measured annulus length ℓ, the rotor speed was raised gradually until circumferential travelling waves developed spontaneously. Long settling times were allowed after each small increment of speed, and thus repeatable, presumably accurate estimates were obtained for the critical value R_w of R at which the steady motion loses stability. The main object was to find R_w as a function of the aspect ratio Γ, but interesting facts were also found from observations on the frequency ω and azimuthal wave-number m of the travelling waves at their inception.

Further, with the same care to allow adequate settling times, the speed was gradually reduced after travelling waves had become established at constant amplitude, and a second critical speed R'_w at which the wave motion ceased was so estimated. In most cases a definite hysteresis was observed, spanned by the interval between R_w and $R'_w < R_w$.

Results for four modes with comparatively few cells are reproduced in Figure 2. From left to right in the figure, the flows represented are (1) the normal two-cell mode, (2) the three-cell mode, (3) the abnormal four-cell mode, and (4) the normal four-cell mode. The most remarkable feature of the experimental curves of R_w versus Γ is that they are sharply peaked, the maxima of R_w being surprisingly high. For the two normal modes, the maxima are about 18 times the critical value of R for the inception of cells according to the idealized theory, and about 15 times an estimate of R_w that has been made on a similar basis with end effects ignored [15]. The maxima for the two anomalous modes are only a little smaller.

The peaks in the curves are understandable on consideration that the individual cells in each array are first compressed and then stretched relative to an optimal length as Γ is varied through the respective range, but this property appears not to have been noticed previously. In experiments where arrays of many cells are observed in a much longer annulus, variations in Γ are known to be accommodated largely by changes in the lengths of the end cells, and so peaks in $R_w(\Gamma)$ may be considerably less pronounced. But they should still be detectable, provided a particular array

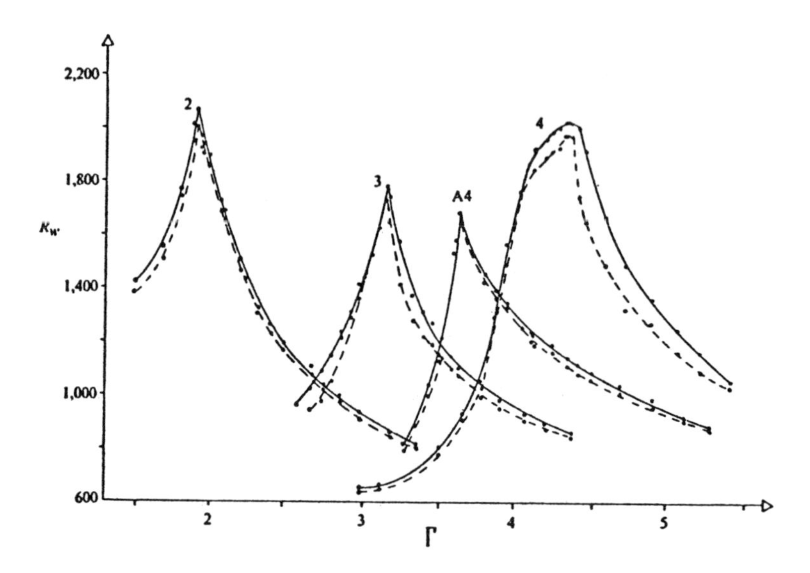

Figure 2. Upper stability limit and hysteresis limit for various cellular modes.

of cells (i.e. a particular N) is tested over a sufficient range of Γ. Cole [16] has reported observations on rapid increases in R_W in step with reductions in Γ below about 10, but his account did not make clear whether the same array of cells had been observed as Γ was varied.

For each of the four modes, it was found that as $R \uparrow R_W$ different wave motions arose in successive ranges of Γ, and corresponding changes were noticeable in the amount of hysteresis (as indicated by the graphs of R_W', drawn dashed in Figure 2). Another remarkable property was that waves in certain ranges of Γ disturbed only part of the cellular flow, the rest of which had no perceptible unsteadiness. For example, the normal four-cell mode developed waves with $m = 2$ if $\Gamma < 4.1$, and then oscillations were apparent on the upper and lower intercellular boundaries but not on the central boundary. On the other hand, in the range 4.1 - 4.34 of Γ, a wave motion with $m = 1$ arose, plainly disturbing the central boundary but not the other two. In the range 4.34 - 4.72 of Γ, a doubly periodic motion disturbing all four cells developed at $R = R_w$, comprising low-frequency waves with $m = 1$ and waves with $m = 5$ whose frequency was approximately (not exactly) five times higher. For $\Gamma > 4.72$, the wave motion at onset was again singly periodic, in this case with $m = 5$.

Whereas such observations indicate that a considerable variety of wave phenomena may arise at the limit of stability for the same cellular flow, and that correspondingly this limit is very sensitive to end effects, another recorded property was found to be remarkably insensitive to them. The dimensionless angular speed $\omega/m\Omega$ of the observed travelling waves was more or less independent of R and m, having a mean value 0.32 among all the measurements of frequency ω. This value is surprisingly close to the value 0.34 found by Coles [17] from measurements in a Taylor apparatus with much larger Γ.

UNSTEADY FLOWS AT HIGH REYNOLDS NUMBERS

The third programme of experiments has been begun in collaboration with T. Mullin, K. Schätzel and E. R. Pike [10]. With the same apparatus as before, operated at high R, a crossed-beam laser rate-correlation method [18], [19] was used to record the fluctuating radial component of velocity at various longitudinal positions within the annulus. From the signals autocorrelation functions were obtained by the computer designed for this purpose at the RSRE, Malvern. Typically, a continuous record of the fluctuations was fed to the computer over a period of about an hour, during which the rotor speed Ω and the temperature of the liquid were automatically held constant, and computed autocorrelation functions revealing considerable information about the observed flows were thus consistently derived. Preliminary findings at Malvern, which will now be summarized, have subsequently been confirmed by more detailed spectral measurements at the University of Kiel, and extensions of the programme are planned.

Figure 3 reproduces representative results at moderate $R > R_w$, for the normal four-cell mode with $\Gamma = 3.68$. The dotted records are measured values of the autocorrelation function $G(\tau) = \langle u(t)u(t + \tau)\rangle$, where $u(t)$ is proportional to the fluctuating component of radial velocity, and the continuous curves are cosines fitted by an automatic least-squares computation. All three records in Figure 3 are normalized by taking the mean-square value $G(0)$ for the first record as the unit of $G(\tau)$, and the delay time τ is given in seconds.

Figure 3(a) was obtained at $R = 11.0R_c$, where $R_c = 71.6$ is an estimate of the critical value of R at which Taylor cells first arise in an infinitely long annulus with $d/r_1 = 0.626$ as in our apparatus. The present vluse $11.0R_c$ is about 30% above R_w for the respective Γ, and the travelling waves here exemplified were still simple-harmonic with $m = 2$. Some harmonic distortion is evident by comparison with the fitted cosine, but this factor has been found generally to be of tenuous veracity, being greatly variable with small positional adjustments of the laser beams. Figures 3(b) and (c) have special interest in showing motions with different spectral features at two positions in the same flow at $R = 20.0R_c$. The measurements were made (b) near the central intercellular interface and (c) in an end cell. The predominant component of the central motion was thus shown to be waves with frequency 1.76 Hz, which had $m = 3$, and that of the end cell to be waves with frequency 0.47 Hz which had $m = 1$. The high-frequency component evidient in Figure 3(c) can be identified as the second harmonic of the central oscillations.

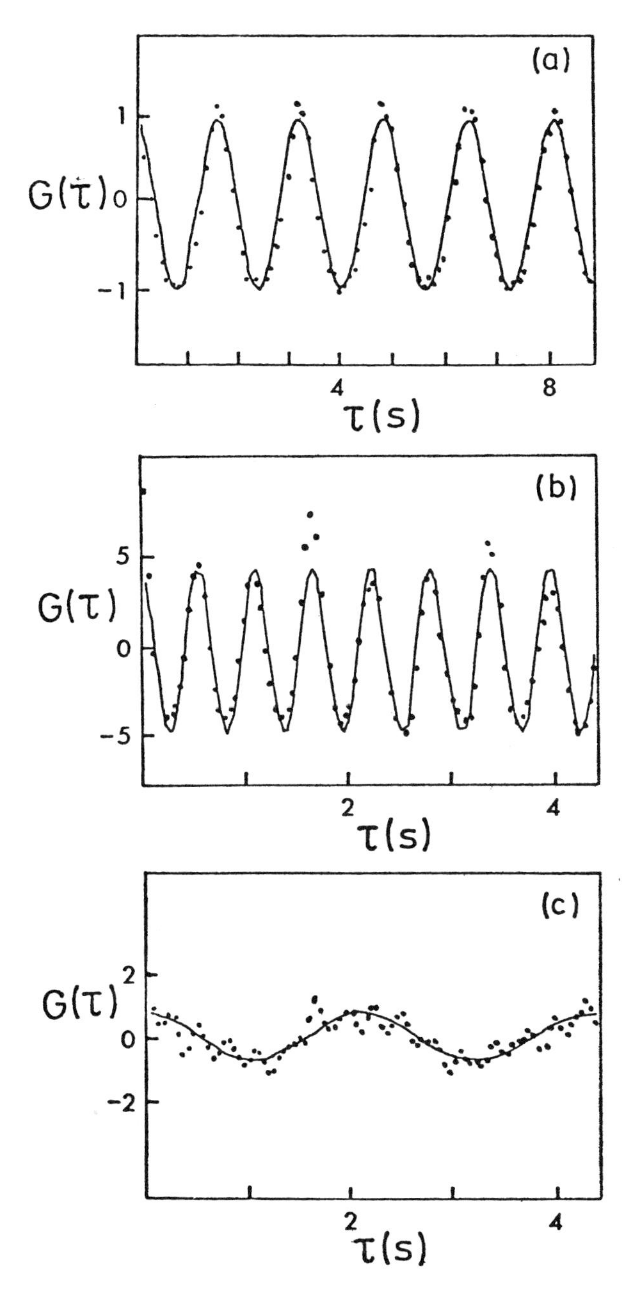

Figure 3. Velocity autocorrelation functions for unsteady four-cell flow: (a) $R = 11.0R_c$, central cells; (b) $R = 20.0R_c$, central cells; (c) $R = 20.0R_c$, end cells.

At higher values of R, more complicated motions were observed, in which as many as six incommensurate, sharp frequencies could be discerned together with some small-scale noise. But another property eventually came into prominence with increasing R, namely the 'sweeping out' of extra periodic components in the central part of the flow, leaving a single moderately energetic periodic component with noise superposed. This behaviour is exemplified in Figure 4, obtained at $R = 33.6R_c$ with $\Gamma = 4.11$. Figure 4(a) shows the motion in the end cells, consisting of noise, as indicated by the initial decay of $G(\tau)$, together with multiply periodic components as indicated by the non-decaying continuation of $G(\tau)$. Figures 5(b) and (c), which present the same $G(\tau)$ on different time-scales, show the central motion having a more orderly form of behaviour. The initial decay of $G(\tau)$ is much more rapid than for the motion at the ends, and the closeness of the cosine (drawn as a dashed line) fitted to the non-decaying continuation of $G(\tau)$ indicates a prominent singly periodic component.

There findings are broadly in accord with those recently reported by Russian experimenters [7], who used a much longer annulus. Respecting the re-emergence of a single sharp spectral component at fairly high R, they also agree with observations by Waldin & Donnelly [6], although diverging from the previous conclusion that a completely disordered state of motion occurs in a range of R below the range of re-emergent order. The present findings are notably different from others recently put on record, for example by Fenstermacher _et al_. [5], where with increasing R an abrupt transition to a "chaotic flow" is judged to occur and for higher R the velocity fluctuations remain completely uncorrelated over long times. In our apparatus R has to be raised to values very much higher than the range exemplified here (including $R = 33.6R_c$) before completely disordered signals are given by the laser anemometer.

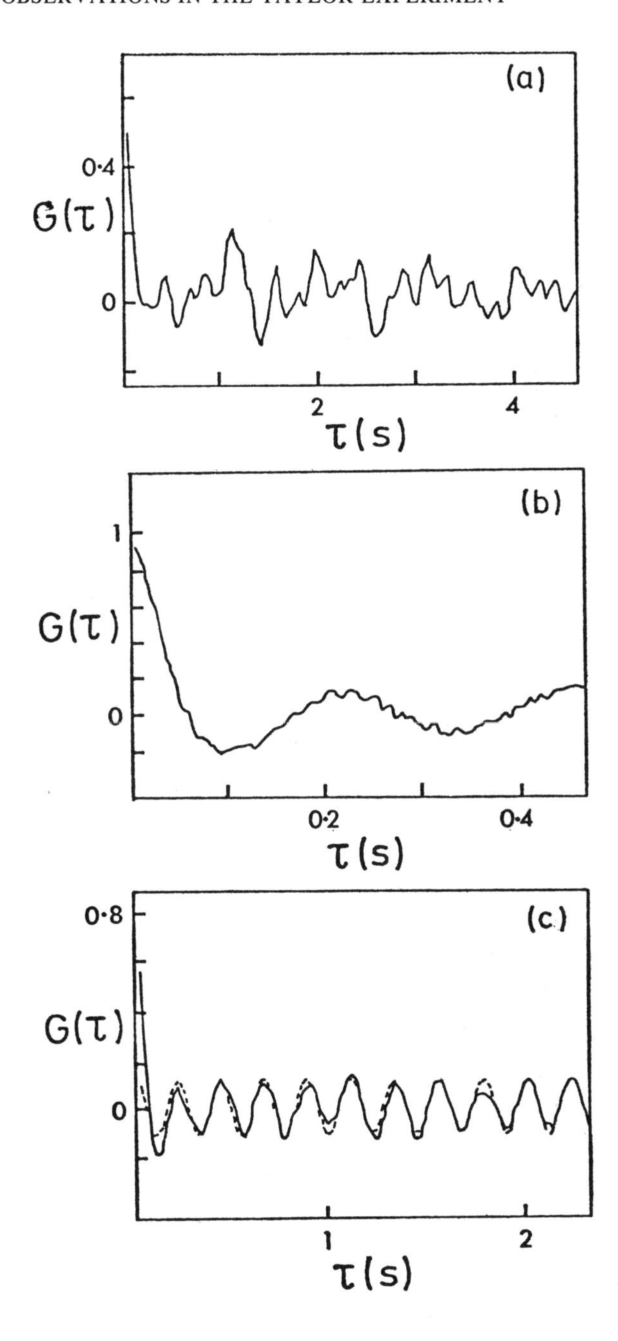

Figure 4. Velocity autocorrelation functions at $R = 33.6R_c$: (a) end cells; (b) and (c) central cells.

CONCLUSION

The phenomena observed in the Taylor experiment with a comparatively short annulus are very complex, but careful experiments over the last few years have put them into satisfactory order. Even for the multiple steady flows that have been investigated, constructive theoretical results adequate for a sure comparison are not yet available, and so interpretations must still rely largely on abstract mathematics. Nevertheless, the robust qualitative theory applicable to the realistic hydrodynamical problem has provided explanations for all the various steady-flow properties observed, including those mentioned here and others detailed in references 8 and 11. As regards the time-dependent phenomena observed at higher R, the variety of realizable behaviour is prodigious, which situation was already vividly demonstrated by the work of Coles [17]; and the prospects for a definite theoretical account still appear remote. New insights have been suggested by analogies with the better-developed theory of dynamical systems with finite freedom, and the circle of ideas concerning strange-attractor behaviour is particularly alluring as a possible means of progress. But its relevance has yet to be demonstrated *a priori* from the Navier-Stokes equations, and so for the time being the latter stages of development of time-dependent behaviour in the Tarylor experiment remain largely mysterious.

Our own experimental studies have led us to a cautious viewpoint regarding the interpretation of observed events in the Taylor experiment. Whereas the use of an apparatus covering a range of comparatively small values of aspect ratio Γ has provided good control over the variety of phenomena enabling information about the complete solution set to be built up systematically, the picture so established is considerably more complicated than a less systematic appraisal might suggest. The realities of the Taylor experiment include many strong effects incapable of description by simpler theories that ignore the presence of the end walls, and the sensitivity of observed properties to parameters other than R, in particular to Γ, is a fact that demands incorporation in any fully rational account. The multiplicity and complex parametric dependency of the solution set cannot be believed to lessen when Γ is greater than the small values exemplified here. Accordingly, the complicated structure underlying the easier observations possible for larger Γ must be expected, in principle, to remain a source of some interpretive obscurity until a full account of the realistic hydrodynamic problem is accomplished.

Finally, a general comment is ventured regarding the status of rotary Couette flow as a prototype of transition to turbulence. It seems reasonable to suggest that because of its invariance under the rotation group, this flow has too much symmetry to be widely representative of transition in practice, notwithstanding its undeniable intrinsic interest. The properties of temporal orderliness observed at moderate and high R plainly owe much to symmetry, for they could certainly be suppressed by, for instance, making the cylinders sufficiently eccentric. Moreover, they have little in common with many other familiar cases of incipient turbulence, for example in pipe flow, in boundary layers and in convection plumes. A case to which, though superficially different, Couette flow seems close in fundamental type is the flow past a circular cylinder perpendicular to the oncoming stream. Here the oscillatory motion developed at moderate R, generating a Kármán vortex street in the wake, is well known to remain a dominant feature of the turbulent motion at high R. (This phenomenon accounts, of course, for the singing of telephone wires in the wind. Minute movements of the wire tend to correlate vortex shedding along its length, so enhancing aerodynamic-sound generation.) The comparison with this case is mainly suggested by the observed phenomena, but a comparable symmetry can be identified in the invariance of the flow to lateral reflection.

REFERENCES

1. E. A. Kusnetsov, V. S. Lvov, Yu. E. Nesterikhin, Y. F. Shmojlov, V. S. Sobolve, M. D. Spetor, S. A. Timokhin, E. N. Utkin & Yu. G. Vasilenko, Inst. Automation & Electrometry, Novosibirsk, Preprint No. 58 (1977).
2. J. T. Stuart, Art. in Applications of Bifurcation Theory (ed. P. H. Rabinowitz), p. 127. (Academic Press 1977).
3. R. C. DiPrima, Transitions in Flow Between Rotating Concentric Cylinders, Proceedings of Symposium on Transition and Turbulence, Mathematics Research Center, University of Wisconsin-Madison, October 1980.
4. T. B. Benjamin, Proc. Symp. on Dynamical Systems, Stability and Turbulence, Univ. of Warwick, 1980 (to appear).
5. P. R. Fenstermacher, H. L. Swinney & J. P. Gollub, J. Fluid Mech. 94, 103 (1979).
6. R. W. Walden & R. J. Donnelly, Phys. Rev. Lett. 42, 301 (1979).
7. E. A. Kusnetsov, V. S. L'vov, A. A. Predtechenskii, V. S. Sobolev and E. N. Utkin, Pis'ma Zh. Eksp. Teor. Fiz. (JETP Lett.) 30, 207 (1979).
8. T. B. Benjamin & T. Mullin, Phil. Trans. Roy. Soc. A (submitted 1980).
9. T. Mullin & T. B. Benjamin, Nature 288, 567 (1980).
10. T. Mullin, T. B. Benjamin, K. Schätzel & E. R. Pike, Phys. Rev. Lett. (submitted 1980).
11. T. B. Benjamin, Proc. Roy. Soc. A 359, 27 (1978).
12. T. B. Benjamin, Math. Proc. Camb. Phil. Soc. 79, 373 (1976).
13. T. B. Benjamin, Proc. Roy. Soc. A 359, 1 (1978).
14. T. B. Benjamin, Art. in Contemporary Developments in Continuum Mechanics and Partial Differential Equations (ed. G. M. de la Penha & L. A. J. Medeiros). Mathematical Studies 30 (1978). (North-Holland).

15. P. M. Eagles, J. Fluid Mech. 49, 529 (1971).
16. J. A. Cole, J. Fluid Mech. 75, 1 (1976).
17. D. Coles, J. Fluid Mech. 21, 385 (1965).
18. R. Vehrenkamp, K. Schätzel, G. Pfister, B. S. Fedders & E. O. Schultz-Dubois, Physica Scripta 19, 379 (1979).
19. J. C. Erdmann & R. P. Gellert, Physica Scripta 19, 396 (1979).

Mathematical Institute
24-29 St. Giles
Oxford OX1 3LB
England

Transition to Turbulence in Thermal Convection with and without Rotation

F. H. Busse

1. INTRODUCTION

Among the fluid systems that exhibit the onset of turbulence those in which the transition occurs gradually are of special interest to theoreticians since the steps in which a system acquires new degrees of complexity can be studied individually. Thermal convection in a layer heated from below and Taylor vortices between differentially rotating coaxial cylinders are the classical examples of fluid systems that exhibit a "slow" onset of turbulence. Much of the recent experimental and theoretical research on the onset of turbulence has been focussed on these two cases. While the observed phenomena are similar in the two systems, there are important differences. A fluid layer heated from below is horizontally isotropic and homogenous at least in the limit when a layer of infinite extent is approached. This causes an infinite degeneracy of the solutions even at the onset of convection. The physical conditions on flow between rotating cylinders, on the other hand, are anisotropic and the onset of Taylor vortices can thus be described by a simple bifurcation from the basic state. In the case of convection this behavior can be modeled by using layers in which height and width are comparable. This permits a removal of the degeneracy to an experimentally relevant

ISBN 0-12-493240-1

extent. The phenomena observed by Gollub and coworkers [12, 13] in small aspect ratio convection layers are indeed rather similar to those observed in Taylor vortex flow [14,20,9]. This review will first briefly describe the phenomena occurring in large aspect ratio and in small aspect ratio convection layers and then focus attention on a recently studied special case of turbulence occuring in a rotating layer of convection.

Mathematical models for the description of the onset of turbulence have usually been developed on the basis of the theory of ordinary differential equations. In the Landau-Hopf picture the transition to turbulence is modeled by a sequence of bifurcations leading to increasingly complex forms of fluid motions as is graphically indicated in figure 1.

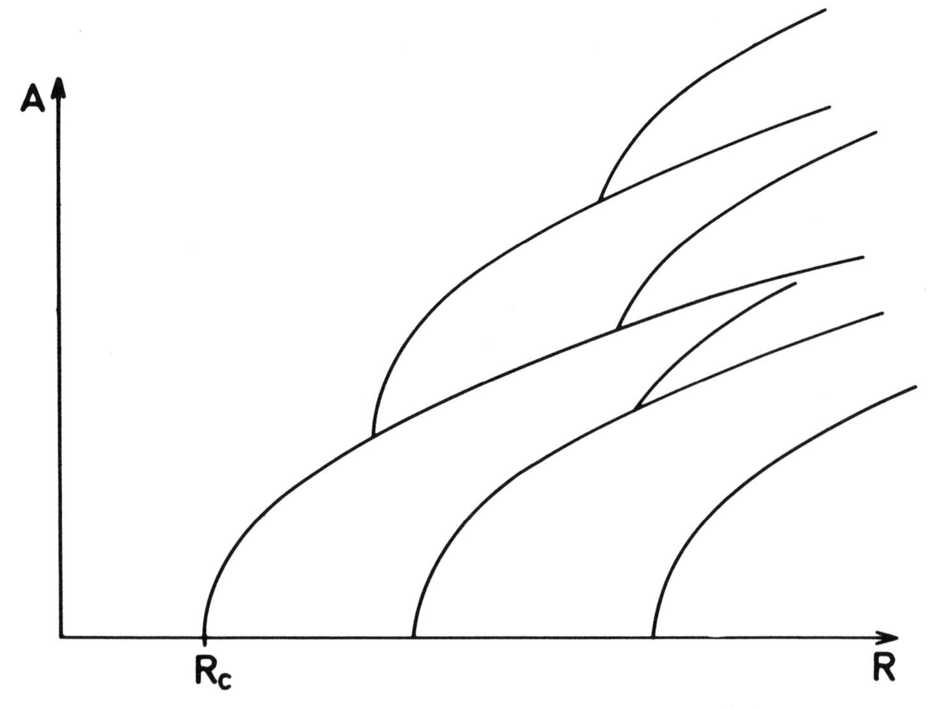

Fig.1. Landau-Hopf picture of transition to turbulence by subsequent bifurcations.

According to the picture of Ruelle and Takens [18], on the other hand, chaotic forms of motions must be expected typically at the fourth bifurcation as is schematically shown in figure 2. Another track leading to chaotic behavior at a finite value R_r of the relevant parameter R is the Feigenbaum [10] sequence of an infinite number of period doubling bifurcations with R_r as point of accumulation. The appearance of strange attractors in the phase space of solutions of ordinary differential equations of higher than second order has led to an intensive effort of research in this area. It will not be attempted here to review the various classes of strange attractors and their relevance to experimental observations. Instead a new model of a peculiar type of turbulence will be presented in section 4 which can be described best as a "statistical limit cycle" and which seems to be appropriate for the interpretation of the phenomena in a rotating convection layer.

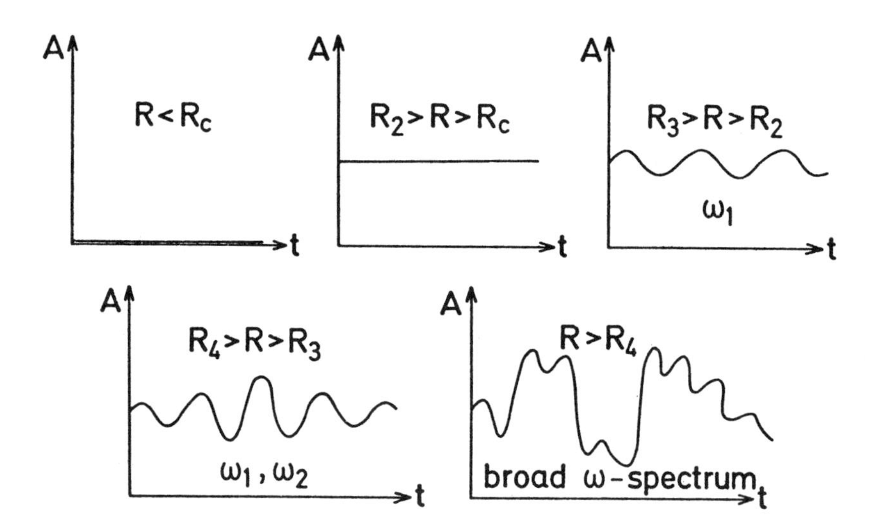

Fig.2. Ruelle-Takens picture of transition to chaotic motion at fourth bifurcation.

It must be realized, however, that all models based on ordinary differential equations are limited in describing the infinite degrees of freedom in a fluid system. They work best for rather confined fluids such as a low aspect ratio convection layer where all modes of motion, except for a few selected ones, are highly damped. Large aspect ratio layers may exhibit different phenomena as is apparent from the experiments of Ahlers and Behringer [1].

2. LARGE ASPECT RATIO CONVECTION LAYERS

When a horizontal layer of a fluid is heated from below and the critical temperature difference across the layer is exceeded, convective motions set in independently in different parts of the layer provided the width to height ratio is sufficiently large, say of the order 10^2. Soon afterwards convection occurs in the regions between the initial patches and the amplitude of convective motion becomes approximately uniform throughout the horizontal extent of the layer. If the fluid satisfies the Boussinesq approximation, i.e. the material properties are constant except for a small linear temperature dependence of the density, roll like motions predominate in the pattern of convection that emerges. But because of the random orientations of the growing initial disturbances the two-dimensional steady solution preferred according to the theoretical analysis [19] can only locally be realized but not globally throughout the layer. The random arrangement of patches of convection rolls exhibits a particular aspect of turbulence. Because of the infinite manifold of existing solutions random initial conditions generate a randomness of the realized flow. The observed macroscopic randomness of the convection pattern thus reflects the randomness of the microscopic disturbances at the time when the critical temperature difference was exceeded. The onset of the convection instability acts, one may say, as a highly selective amplifier of microscopic noise at the initial time.

Since the pattern of convection generated from random initial conditions changes only rather slowly, one usually does not call it turbulent. The transition to turbulence in convection is associated with the appearence of complex three-dimensional structures and aperiodic oscillations. In order to identify experimentally the physical processes which induce these turbulent properties, it is necessary to eliminate the above mentioned randomness generated by uncontrolled initial conditions. This can easily be done by introducing small temperature disturbances of prescribed wavelength in the convection layer before raising the temperature difference to its critical value. In that case convection sets in in the form of nearly exact two-dimensional rolls and the parameters of instabilities of steady rolls can be measured quantitatively. Typically the three-dimensional form of convection replacing the two-dimensional rolls still exhibits a high degree of regularity. By continuing the experiment into the range of higher temperature differences or higher Rayleigh numbers, as the corresponding dimensionless parameter is called, new instabilities can be discovered and their physical properties determined quantitatively. The method of controlled initial conditions is thus capable of isolating the discrete transitions leading to turbulent convection which are observable but not clearly distinguishable in an experiment with random initial conditions.

As an example we consider the experimental observations shown in figure 3 of four subsequent bifurcations in a fluid of moderate Prandtl number heated from below. Starting with two-dimensional convection rolls as the first bifurcating solution of the problem the transition to steady bimodal convection occurs when the Rayleigh number approaches a value of the order $2 \cdot 10^4$. The third bifurcation to oscillatory bimodal convection becomes noticeable a little above twice that value. At a slightly higher Rayleigh number oscillatory bimodal convection experiences a subharmonic instability in that several cells of the bimodal pattern are collected into the new

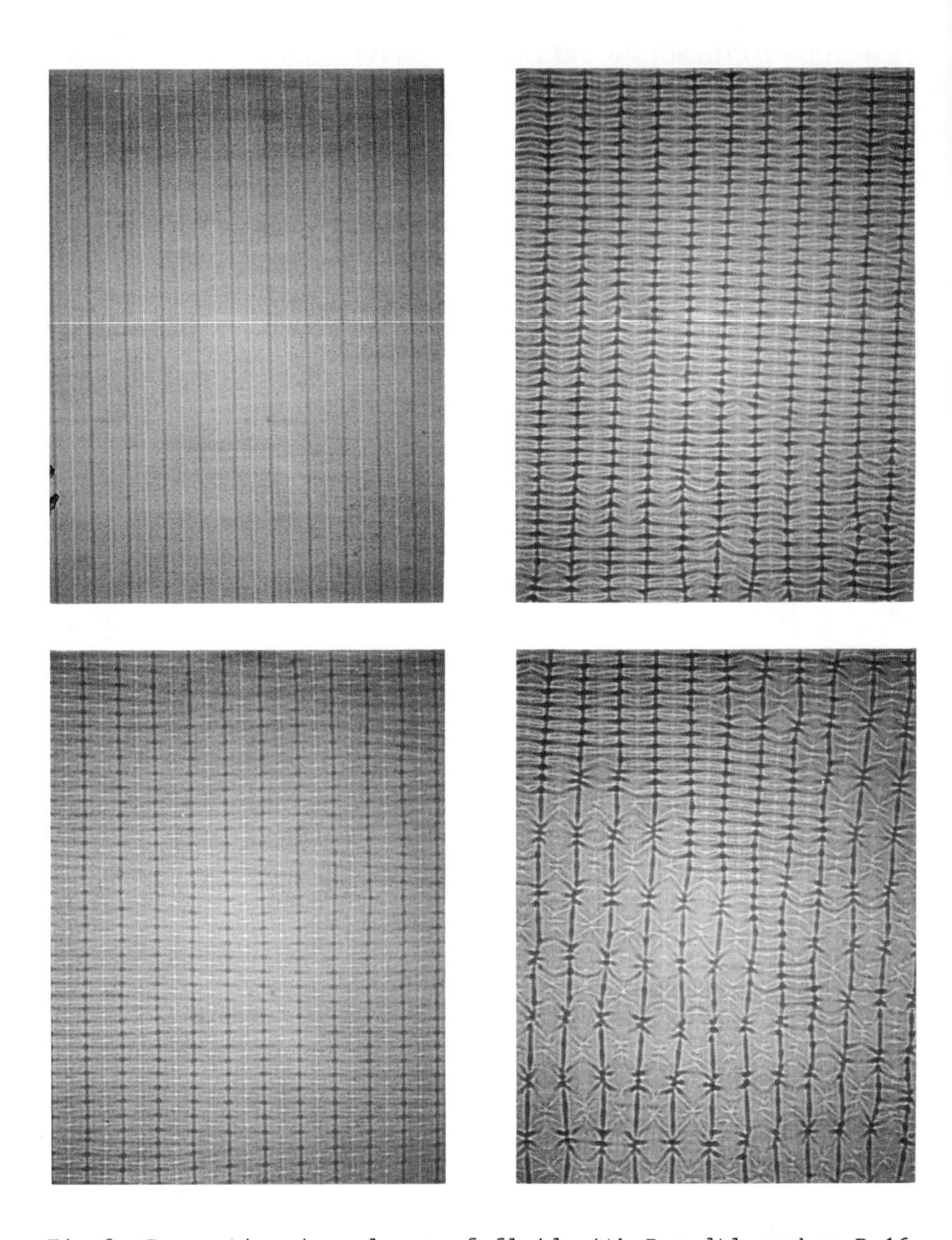

Fig.3. Convection in a layer of fluid with Prandtl number P=16 (1cst silicone oil). The Rayleigh number increases from about $2 \cdot 10^4$ (upper left) to about $6 \cdot 10^4$ (lower right). For details see [7]. The pictures show convection rolls, bimodal convection (lower left), oscillatory bimodal convection (upper right) in which the longer sides of cells move forth and back and, finally, the onset of spoke pattern convection.

basic periodicity interval of spoke pattern convection. The large scale structure of this form of convection remains essentially steady, while the small scale "spokes" exhibit a fluctuating time dependence with a spectral peak close to the frequenecy of oscillatory bimodal convection. Since the bifurcations involve changes in both the spatial as well as the time dependence of convection, simple models are not feasible in this case. But the mechanism of instability can be understood qualitatively on the basis of the known instabilities of convection rolls.

The theoretical analysis of the transitions leading to turbulent convection has mostly been restricted to the description of secondary bifurcations. Since steady two-dimensional solutions, describing convection in the form of rolls, represent the only stable steady solution at the critical value of the Rayleigh number [19], it is a task of only moderate computational expense to generate accurate approximations for the experimentally realisable solution of the basic nonlinear equations. By superimposing arbitrary three-dimensional disturbances of infinitesimal amplitude onto the steady solution the stability of the latter can be tested. The disturbance with the highest real part of the growthrate at the point where the real part vanishes determines the limit of the stability region of the steady rolls. Since the two-dimensional solution depends on the Rayleigh number R and the Prandtl number P as external parameters and the horizontal wave number α as intrinsic parameters, the region of stable two-dimensional solutions is bounded by a tube like surface in the three-dimensional parameter space. As shown in figure 4 the surface is formed by different instabilities which can be catalogued according to various properties. Instabilities restricting the domain of stable rolls in the dimension of the wavenumber α typically accomplish a transition to rolls of different wavelength as can be confirmed by comparison with experimental observations. More interesting with respect to the problem of

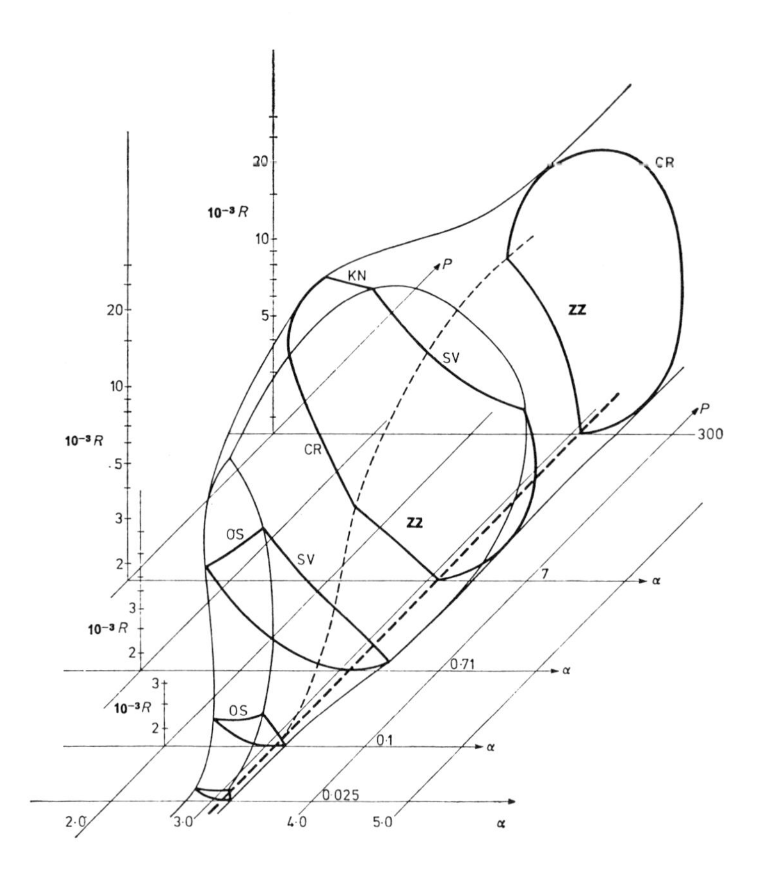

Fig.4. Region of stable two-dimensional convection rolls in the R - α - P parameter space. This graph first published in [2] is a composite picture of results obtained in earlier work by R.M. Clever and the author. The lettering of the stability surface indicates the mechanism of instability; OS: oscillatory instability, CR: crossroll instability, SV: skewed varicose instability, KN: knot instability, ZZ: zig zag instability. The stability boundary for P = 300 actually represents the computations for P = ∞ which are expected to give a good approximation for P = 300.

the transition to turbulence are the pattern changing instabilities which typically bound the domain of stable rolls towards high Rayleigh numbers. Among these instabilities the oscillatory instability [3,8] is of special importance since it introduces a time dependence of the convection flow. For a more detailed discussion of the various instabilities the reader is referred to a recent review [2] and to the additional paper [5]. Here we like to emphasize that the various mechanisms of instabilities of rolls can be used for the interpretations of the instabilities of more complicated convection patterns. The fact, for example, that the transition to oscillatory convection is preceded by the transition to steady three-dimensional bimodal convection does not prevent the occurrence of oscillations at roughly the same Rayleigh number as would have been predicted on the basis of the stability analysis of rolls. The oscillations of bimodal convection correspond to standing waves, of course, since the three-dimensional structure of bimodal convection prevents propagating waves which are the typical form of oscillations in the case of rolls. But the results for the latter can be used to predict the frequency of oscillatory bimodal convection.

3. LOW ASPECT RATIO CONVECTION LAYERS

The property of degeneracy which characterizes convection in large aspect ratio layers disappears when the width of the layer becomes comparable with its depth within a factor of about 10. The side walls of such convection "box" exert a sufficiently strong constraint to reduce the number of realized convection flows to a few cases with distinct structure, the actual number increasing with the aspect ratio. Although a finite domain of a fluid still exhibits infinite degrees of freedom, in practice only a finite number can be excited because of the strong viscous damping of modes with short length scales. Because of the limited number of degrees of freedom that must be considered, mathematical models based on systems of ordinary differential equations describing the time dependence of a few selected modes are capable of describing experimentally observed phenomena in a convection box.

Since the advent of laser Doppler velocimetry it has become feasible to obtain detailed measurements of convection flows without any noticeable interference. Gollub et al.[13] and Gollub and Benson [12] have reported on the sequence of transitions starting with the onset of steady convection followed by the onset of periodic oscillations, the appearance of a second basic frequency of oscillation and, finally, the onset of a chaotic time dependence characterized by a broad band noise. Spectral analysis is essential to distinguish the different regimes, since the recorded velocity amplitudes in the quasiperiodic regime of oscillations do not seem to differ much from those recorded in the aperiodic regime as shown in figure 5.

Phase locking of the originally independent modes of oscillation occurs typically at Rayleigh numbers slightly less than the value at which the broad band noise develops. The latter transition indicates the inability of the system to achieve permanent phase locking. The inspection of the velocity record shows that periodic motions occur in segments of the record connected by aperiodic excursions. But the particular behavior shown in figure 5 represents only one of several possible tracks in the development of turbulence in a convection box. As Gollub [11] has emphasized, slight changes in the Prandtl number or the aspect ratio of the convection box can lead to qualitatively different routes in the development of turbulence. Among those routes the period doubling of the oscillations is of special interest because of its correspondence to the theory of Feigenbaum [10]. For details on this and other experimental observations we refer to the above mentioned papers by Gollub and coworkers.

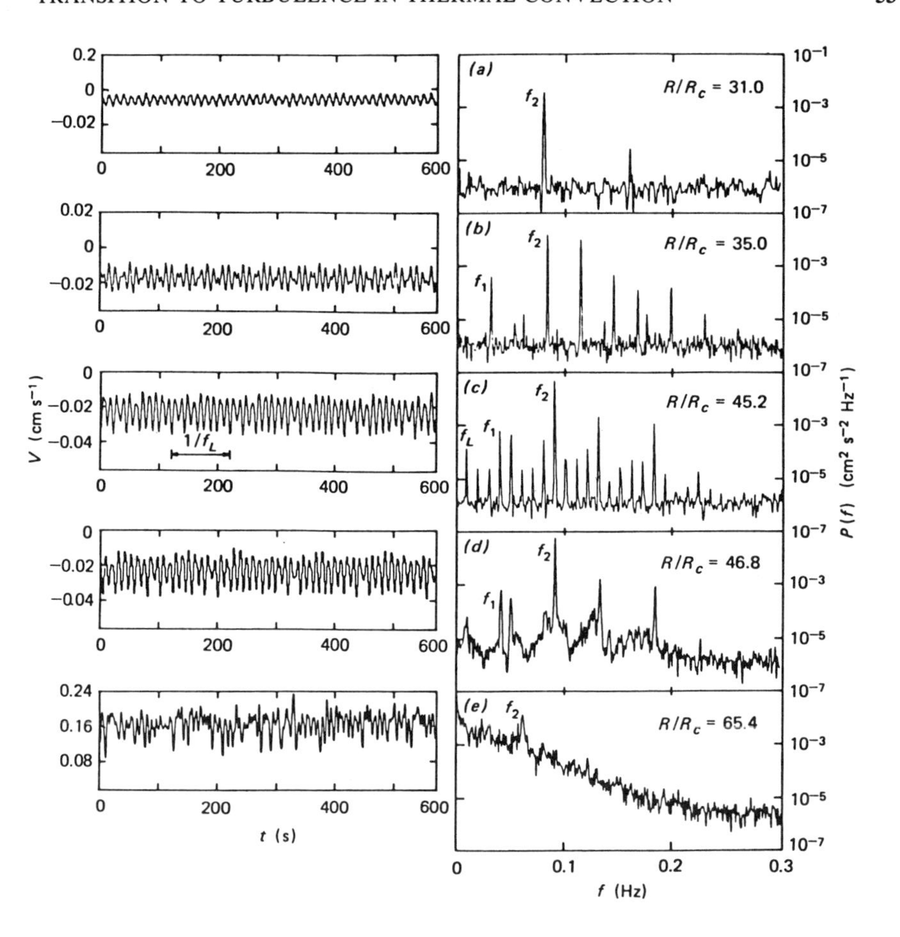

Fig. 5. Velocity records and power spectra showing the sequence of instabilities leading to a periodic flow after Gollub and Benson [12]. The sequence of spectra shows (a) a periodic state with a single peak and its harmonics; (b) a quasiperiodic state with 2 incommensurate frequencies f_1 and f_2 and their linear combinations; (c) phase locking at the integer ratio $f_2/f_1=9/4$; (d) an aperiodic state with relatively sharp peaks above a broadband spectrum; and (e) a strongly aperiodic state.

4. RANDOM CONVECTION IN A ROTATING LAYER

Convection in a horizontal layer heated from below and rotating about a vertical axis exhibits a time dependence with statistical properties even in the limit of small amplitudes of the convective motion. It thus represents a rather unique example in that aspects of turbulence can be investigated in a weakly nonlinear system. In the non-rotating case two-dimensional convection in the form of rolls is the only stable form of convection. The extension of the small amplitude perturbation analysis of [19] by Küppers and Lortz [16] produced the result that even the roll solution becomes unstable when the rotation rate exceeds a critical value. The Küppers and Lortz instability assumes the form of roll disturbances with an axis inclined by an angle of about 60° with respect to the axis of the given steady rolls. But the instability cannot lead to a new steady or oscillatory form of convection since all steady solutions are unstable and oscillatory solutions do not exist in the neighborhood of the critical Rayleigh number when the Prandtl number exceeds unity as in most experimental situations.

To resolve this paradoxical problem the initial value problem must be solved [4]. Since the orientations of growing and decaying roll solutions differ by nearly 60°, a minimum of three modes is sufficient to give an approximate description. Using the z-coordinate in the vertical direction, the vertical component of the velocity can be written in the form

$$u_z = \left[\hat{g}(t) \cos \underset{\sim}{k}_1 \cdot r + \hat{g}_2(t) \cos \underset{\sim}{k}_2 \cdot r + \hat{g}_3(t) \cos \underset{\sim}{k}_3 \cdot r\right] f(z) + \ldots \quad (1)$$

where higher order terms have not been given explicitly. The three horizontal $\underset{\sim}{k}$-vectors are all of equal length and satisfy the relationship

$$\underset{\sim}{k}_1 + \underset{\sim}{k}_2 + \underset{\sim}{k}_3 = 0 \ .$$

By introducing $g_n(t) = \gamma|\hat{g}_n(t)|^2$ and by using appropriate transformations [4] the equations for the amplitudes $g_n(t)$ can be written in the form

$$\dot{g}_1 = g_1(1 - g_1 - \alpha g_2 - \beta g_3) \quad (2a)$$

$$\dot{g}_2 = g_2(1 - g_2 - \alpha g_3 - \beta g_1) \quad (2b)$$

$$\dot{g}_3 = g_3(1 - g_3 - \alpha g_1 - \beta g_2) \quad (2c).$$

This system of equations has been investigated in the context of population biology by May and Leonard [17]. In the present context the region of the parameter space $\alpha+\beta > 2$ with $\alpha < 1$ or $\beta < 1$ is of interest in which all fixpoints of the system of equations (2) are unstable. In this case the solution of (2) approaches rapidly the plane $g_1 + g_2 + g_3 = 1$ and then wanders from the neighborhood of one of the fixpoints $g_n=1$, $g_m=0$ for $m \neq n$, to that of another in a sequence depending on the sign of $\alpha - 1$. The typical solution of the initial value problem, sketched in figure 6, shows that the time spent in the neighborhood of anyone of the three fixpoints increases monotonically and so does the period of the entire cycle. This inhomogeneous time dependence cannot correspond to a physically realistic system. The experimental noise alone prevents the decay of the amplitudes g_n to the arbitrary small values assumed by the solutions of (2).

In the actual experiment [15,6] the phases of the solutions of (2) differ at different places of the rotating convecton layer. Thus neighboring patches of rolls, rather than the experimental noise, provide a minimum level of the disturbance amplitudes and thus prevent a monotonic increase of the cycle period. The influence of the neighboring patches increases with decreasing size of the given patch of rolls. But the size L of a patch is determined by the balance between the time of horizontal diffusion L^2/κ where κ is the thermal diffusivity and the time of growth of the instability. The latter is roughly proportional to $(R-R_c)^{-1} \cdot (T-T_c)^{-1}$, where R_c is the critical

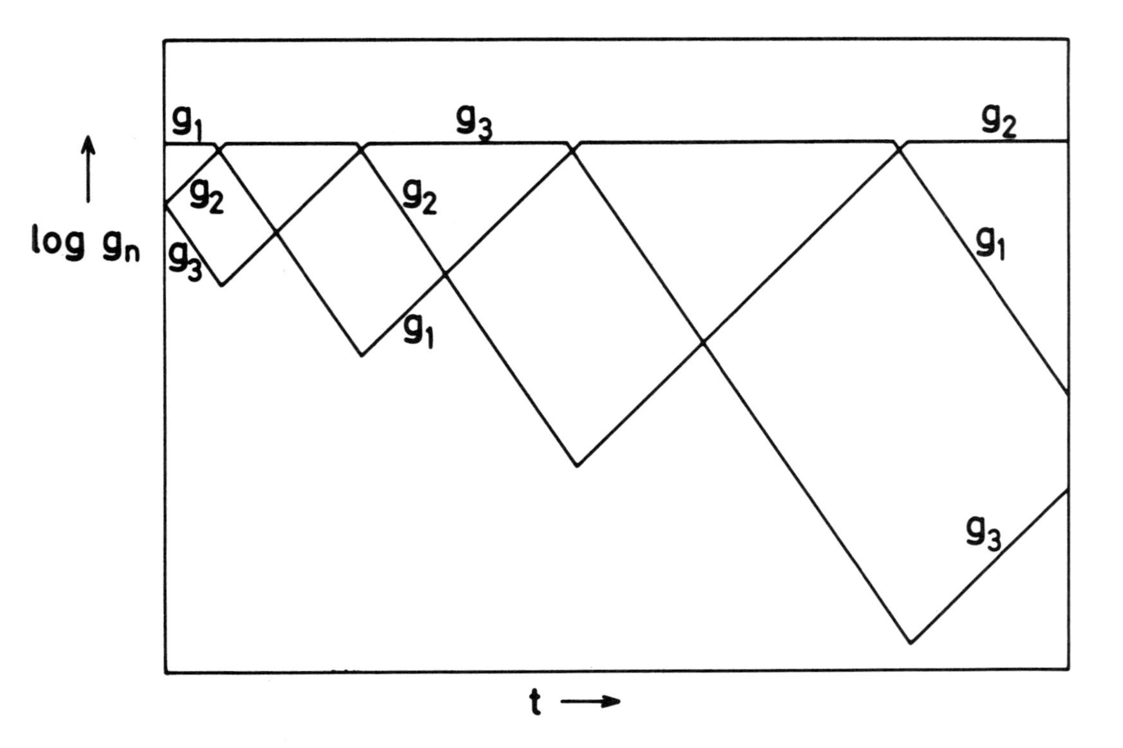

Fig.6. Sketch of a solution of equations (2).

value of the Rayleigh number R for the onset of convection, while T_c is the critical Taylor number for the onset of the Küppers - Lortz instability. We thus arrive at the qualitative description of the physical system sketched in figure 7. Since the energy of convection given by $g_1+g_2+g_3$ remains essentially constant after a short initial period the attention can be restricted to the plane $g_1+g_2+g_3=1$. The solution of equations (2) indicated in figure 7a is not physically realisable even in the idealized limit $R=R_c$, $T=T_c$, since effects of experimental noise or even thermodynamic fluctuations will restrict the validity of equations (2). But for relatively small values of $R-R_c$ and $T-T_c$ those effects can be neglected in comparison with the influence of the neighboring patches on the time dependence of convection at a given location in the convection layer. This effect leads to the limit cycle property of the solution indicated in figure 7b. But since the effect is stochastic in nature, the limit cycle has only a statistical meaning. The period will fluctuate from one cycle to the next. As the growthrate of the Küppers-Lortz instability increases with increasing values of $R-R_c$ and $T-T_c$, the statistical limit cycle shrinks until it represents only a small fluctuation about the steady solution

$$g_n = \frac{1}{3} \qquad \text{for } n = 1,2,3 \qquad (3).$$

This solution describes convection in the form of hexagonal cells. Such cells have indeed been observed in the experiment.

5. CONCLUSION

The fact that typical properties of turbulent fluid systems can be observed near the critical value R_c for the onset of convection in a rotating layer and, to some extent, also in a nonrotating convection layer originates primarily from the degeneracy of the solutions of the basic bifurcation problem. The availability of an infinite manifold of solutions corresponding to all possible horizontal orientations of convection rolls provides the basis for the realisation of random

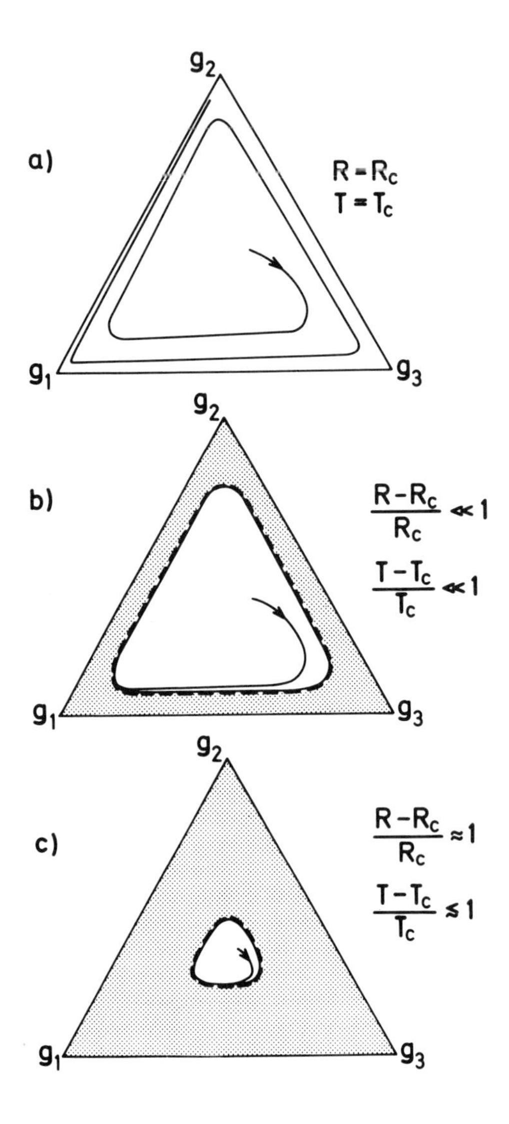

Fig.7. Sketch of the statistical limit cycle (dashed line) for different values of the parameters.

patterns of convection. Although the infinite degeneracy is a rather unique property of convection in layers of infinite extent, it is also typical for highly turbulent fluid systems. The effects of anisotropy tend to become less important in high Reynolds number flow when a large number of nearly equivalent solutions becomes available.

The concept of a mathematical description of turbulence by a large manifold of solutions, each of which is unstable with respect to some other solution of the manifold, is not a new one. Convection in a rotating layer offers the simplest example of such a model of turbulence with the additional advantage of a weak nonlinearity of the basic equations. This should permit a more detailed theoretical investigation of various aspects of the problem among which the interaction of patches of rolls is of special interest.

REFERENCES

1. Ahlers, G. and R. P. Behringer, Evolution of turbulence from the Rayleigh-Benard Instability, Phys. Rev. Lett. 40, 712 - 716, 1978.
2. Busse, F. H., Nonlinear Properties of Convection, Rep.Progress in Physics 41, 1929 - 1967, 1978.
3. Busse, F. H., The oscillatory instability of convection rolls in a low Prandtl number fluid, J. Fluid Mech., 52, 97 - 112, 1972.
4. Busse, F. H. and R. M. Clever, Nonstationary convection in a rotating system, in Recent Developments in Theoretical and Experimental Fluid Mechanics, (U. Müller, K.G. Roesner, B.Schmidt eds.), Springer, 1979, pp. 376 - 385.
5. Busse, F. H. and R. M. Clever, Instabilities of convection rolls in a fluid of moderate Prandtl number, J. Fluid Mech. 91, 319 - 335, 1979.

6. Busse, F. H. and K. E. Heikes, Convection in a rotating layer: A simple case of turbulence, Science 208, 173-175, 1980.
7. Busse, F. H. and J. A. Whitehead, Oscillatory and collective instabilites in large Prandtl number convection, J. Fluid Mech. 66, 67 - 79, 1974.
8. Clever, R. M. and Busse, F. H., Transition to time dependent convection, J. Fluid Mech., 65, 625 - 645, 1974.
9. Donnelly, R. J., N. Park, R. Shaw and R. W. Walden, Early nonperiodic transitions in Couette flow, Phys. Rev. Lett. 44, 987 - 989, 1980.
10. Feigenbaum, M. J., Quantitative universality for a class of nonlinear transformations, J. Statist. Phys. 19, 25-52, 1978.
11. Gollub, J. P., The onset of trubulence: convection, surface wave and oscillators in: Systems Far from Equilibrium, (Proceedings of the Sitges International School and Symposium on Statistical Mechanics), Ed. by L. Garrido, Lecture Notes in Physics, Springer Verlag, 1980.
12. Gollub, J. P. and S. V. Benson, Many routes to turbulent convection, J. Fluid Mech. 100, 449-470, 1980.
13. Gollub, J. P. S. L. Hulbert, G. M. Dolny and H.L. Swinney, Laser Doppler study of the onset of turbulent convection at low Prandtl number, in Photon Correlation Spectroscopy and Velocimetry, Plenum Press, New York and London, pp. 425 - 39, 1977.
14. Gollub, J. P. and H. L. Swinney, Onset of turbulence in a rotating fluid, PHy. Rev. Lett. 35, 972 - 930, 1975.
15. Heikes, K. E., An experimental study of convection in a rotating layer, Ph. D. Thesis, University of California at Los Angelos, 1979.
16. Küppers, G. and D. Lortz, Transition from laminar convection to thermal turbulence in a rotating fluid layer, J. Fluid Mech. 35, 609 - 620, 1969.
17. May, R.M. and W.J. Leonard, Nonlinear aspects of competition between three species, SIAM J. Appl. Math. 29, 243 - 253, 1975.

18. Ruelle, D. and F. Takens, On the nature of turbulence, Commun. Math. Phys. 20, 167, 1971.
19. Schlüter, A., D. Lortz and F. Busse, On the stability of steady finite amplitude convection, J. Fluid Mech. 23, 129 - 144, 1965.
20. Walden, R.W. and R.J. Donnelly, Reemergent order of chaotic circular Couette flow, Phys. Rev. Lett. 42, 301 - 304, 1979.

Department of Earth and Space Sciences
University of California
Los Angeles, CA 90024

Instability and Turbulence in Jets

J. Laufer

1. INTRODUCTION

Research in turbulence has undergone considerable changes in the last decade. This change was brought about by the discovery of large, spatially coherent structures in fully developed flows.[1,2] As a consequence, the emphasis shifted from the study of the statistical nature of the flow to the deterministic aspects of these structures. Furthermore, it was soon realized that in free turbulent shear layers the initial instability of the flow can have a surprisingly strong influence on its subsequent development; thus, the instability and the subsequent turbulent state of the flow are coupled. The main purpose of this paper is to explore in some detail the nature of this coupling.

The new point of view is affecting changes in the mathematical formulation of the problem as well. In the past the instability and turbulence were treated quite separately and differently. The former was cast into an eigen-value problem in which an unsteady, linear partial differential equation, the well known Orr-Summerfeld equation, was solved; the point of departure of the latter one is the Reynolds equations, a set of time averaged, nonlinear partial differential equations that are incomplete. In discussing some of the recent experimental findings, it will become clear that neither

TRANSITION AND TURBULENCE

ISBN 0-12-493240-1

formulations are adequate for a turbulence theory. At the same time it is hoped that the results of the new trend will suggest new directions for further theoretical developments.

Although the presence of large scale, vortex-like structures have been observed many years ago, their existence at high Reynolds number and their importance to the flow development has been recognized only recently. The main objective here is to study the behavior of these structures in space and time and in particular to understand the dynamics of their motion. Due to the lack of any theoretical guidance, the study is still descriptive and speculative. Nevertheless, a number of unexpected and possible significant results can be already reported.

THE TURBULENT JET

For the purpose of our discussion, a simple flow geometry, the turbulent axisymmetric jet will be chosen that received considerable attention in the past. Some of the general features that emerged from these studies are summarized below.

The flow field may be divided into four regions (Figure 1):

1) The instability region; the unstable shear layer develops oscillations generating periodic vortex rings;
2) The mixing region; the vortex rings interact with each other and thereby spread the vorticity until the center, potential core region disappears;
3) The transition region; the flow relaxes into its final stage of development;
4) The fully developed region, in which self-similarity is reached in the flow.

It is found that the spreading rate does not change with Reynolds number, but is affected by compressibility.

In the past, most of the experimental work concentrated on Region 4. It was realized, however, that measurements of statistical quantities, such as the Reynolds stresses, the spectra or the spatial correlations of the velocity fluctuations are not helpful to a better understanding of the flow

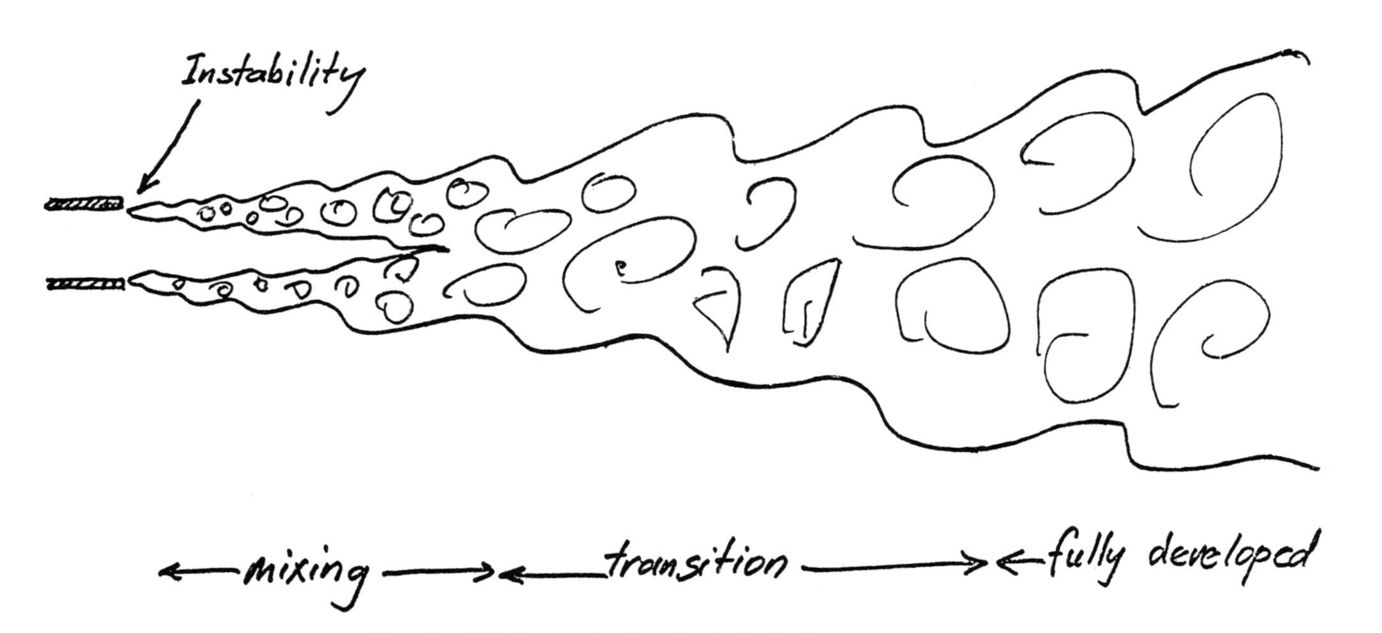

Fig. 1 Development of a Turbulent Jet.

dynamics. For this reason, attention shifted to the first two regions of the flow field with the idea that a study of the behavior of vortex ring like structures would reveal some clues about the flow development.

a) The Impinging Jet

One of the most interesting and revealing studies was recently conducted by Ho and Nosseir in a so-called impinging jet, in a flow in which a turbulent jet impinges against a plane solid surface placed perpendicular to the jet axis.[3] They found that if the surface is located within about six diameters from the nozzle exit and if the jet Mach number is above approximately 0.7, a strong resonance develops in the flow. The intriguing question that arises concerns the mechanism that is responsible for the selection of a particular frequency, the resonance frequency, out of a wide range that exists in a turbulent jet. The problem is a good example to demonstrate the important role the initial instability and the subsequent large scale structures play.

Ho and Nosseir found that the resonance is the result of a feedback loop that may be described as follows (Figure 2): the large scale structures developed in the mixing layer impinge on the plate; because of their spatial coherence they generate a pressure field with a well defined phase reference that propagates upstream and interacts with the initial shear layer; the interaction generates periodic structures that are convected downstream to the plate completing the feedback loop.

Ho and Nosseir have also studied the details of the interaction between the initial vortices of the shear layer and the upstream moving pressure field (Figure 3). The vortex generation frequency is determined by the initial shear layer instability and is generally at least an order of magnitude higher than the resonance frequency. The pressure field forces the shear layer into a contracting-expanding motion as shown in the figure, resulting in an accelerating and decelerating flow that in turn changes the stability characteristics of the layer. In particular, during the deceleration phase the initial vortices roll up into a large one. This

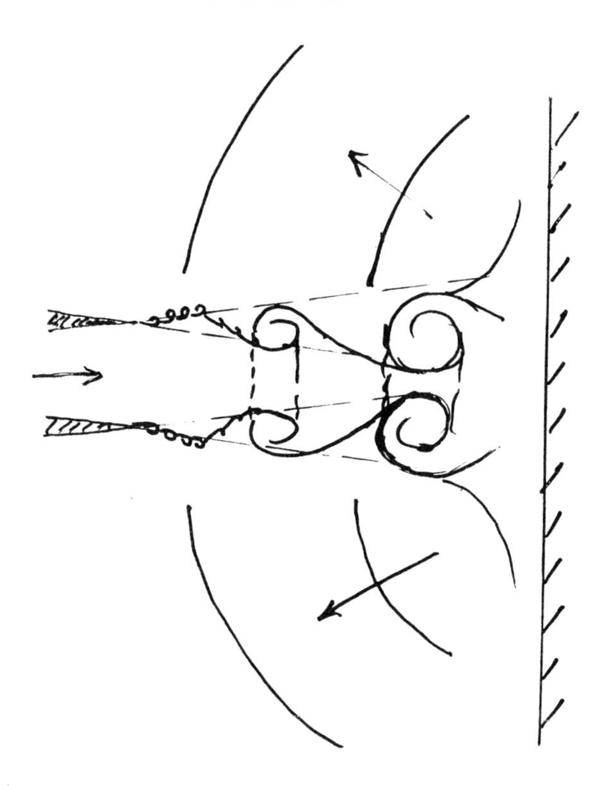

Figure 2 Impinging Jet

process, called collective interaction, thus, insures a phase lock between the large structures and the upstream propagating pressure and in effect decouples the initial instability frequency from the resonance frequency. The resonance frequency is determined by the requirement that the period for the structures to convect from the nozzle to the plate and the pressure perturbation to propagate back to the nozzle should be an integer multiple of the resonance period, $1/f_r$. If x_o is the plate separation distance from the nozzle, V_c and a the convection and acoustic propagation velocities respectively, and N an integer then

$$\frac{x_o}{V_c} + \frac{x_o}{a} = \frac{N}{f_r}$$

Figure 4 confirms the above described physical picture.

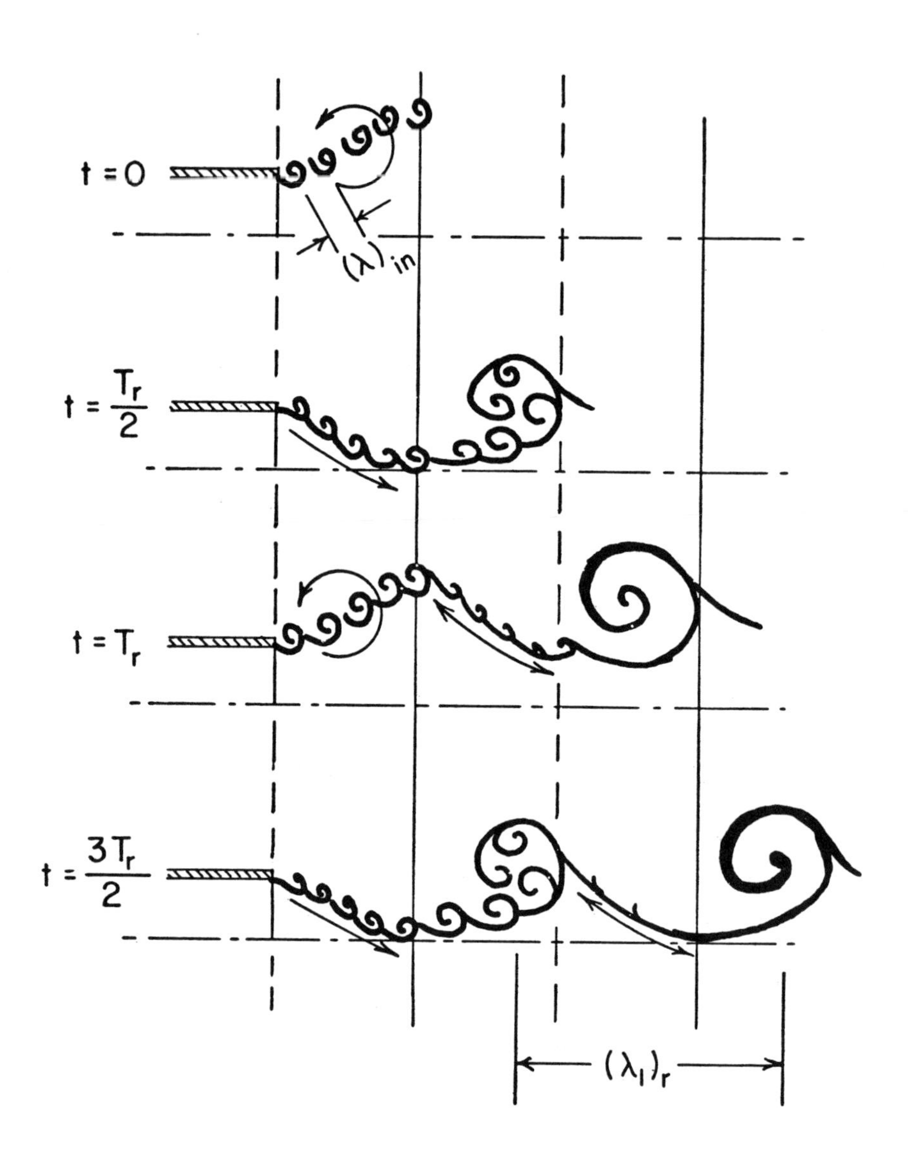

Figure 3 Collective Interaction (Ref. 3)

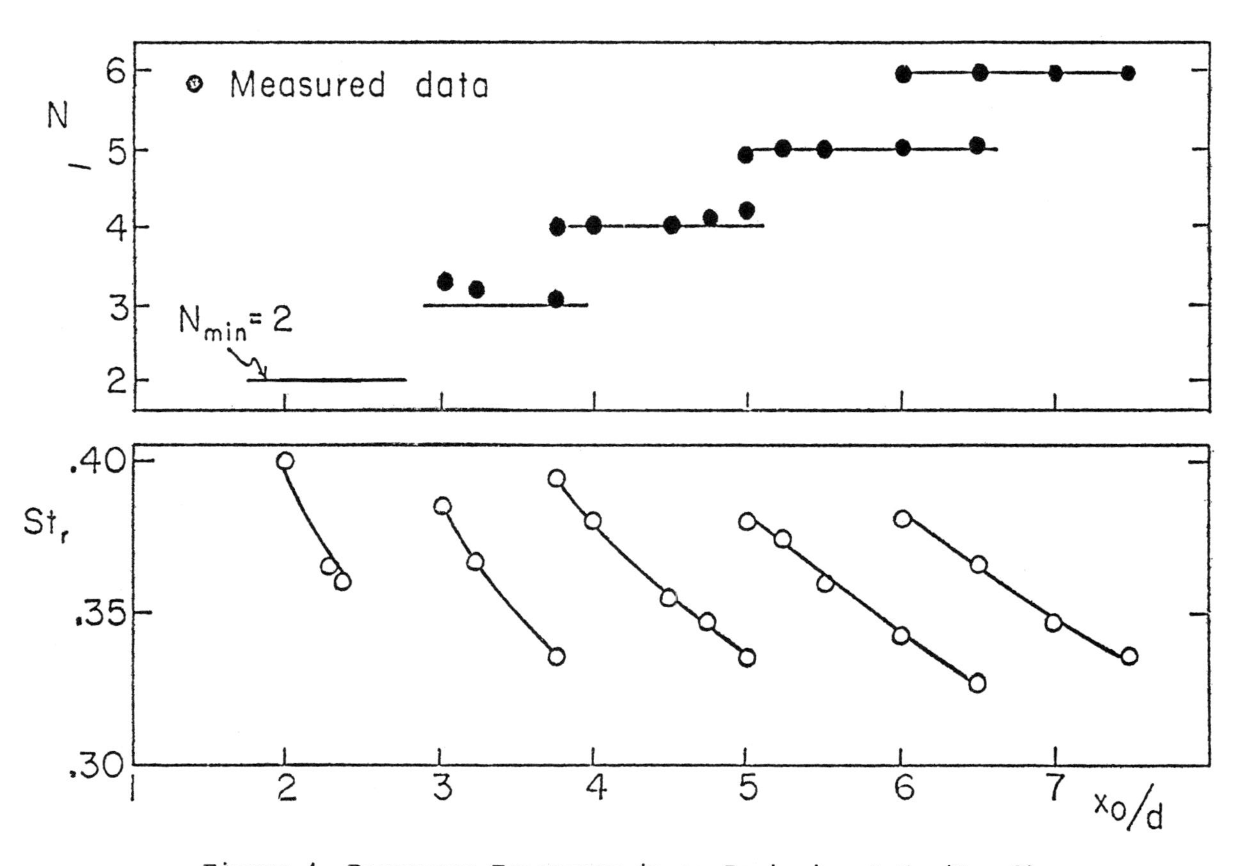

Figure 4 Resonance Frequency in an Impinging Jet (M_j=.9).

The lower portion of the figure shows the conventional way of plotting the nondimensional resonance frequency, $St_r = \frac{f_r D}{V_j}$ versus the separation distance x_o/D where D and V_j are the diameter and jet velocity respectively. Since N can be calculated using the above equation and the measured values of V_c, the upper portion of the figure clearly shows how the resonance frequency jumps from one integer value of N to another depending on the plate separation distance.

b) The Free and Forced Jet

The question naturally arises whether or not a feed-back mechanism is present in a free jet as well. The problem was investigated by Laufer and Monkewitz.[4] They have analyzed the signature of a hot wire placed in the unstable shear layer of a jet very close to the nozzle and found that the amplitude of the unstable oscillations is highly modulated (Figure 5). A spectrum analysis of the demodulated signal clearly shows a strong peak at St = 0.31 (Figure 6). This corresponds to the so-called "preferred mode" of the jet, that is to the passage frequency of the large scale structures near the end of the potential core. This mode is known to occur in the range $.25<St<.5$ in subsonic jets.

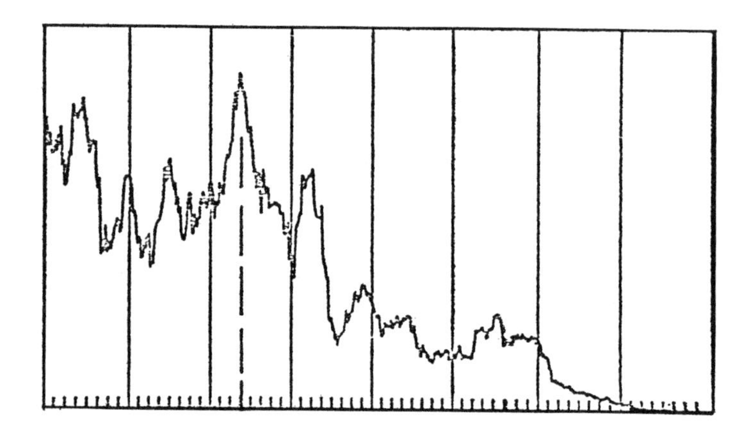

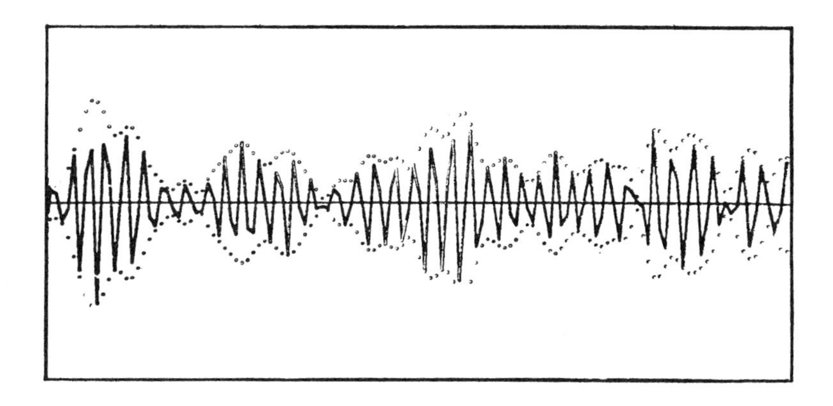

Figure 6 Spectrum of the demodulated signal near the nozzle exit.

The result is interesting in that it points to the fact that large structures located as far as four or five diameters downstream of the nozzle influence the initial shear layer. The idea of a feed-back is an attractive one because it provides a possible mechanism for the vortex interaction, that is, the pairing process and a further examination of it is worthwhile.

Recent numerical simulation studies[5,6] have indicated that in a shear layer the presence of an appropriately phased subharmonic is essential to the pairing process. Accordingly, the vortices in Figure 7 tend to pair if a cross stream perturbation has the indicated phase relationship with them. It is conceivable that in a developing shear layer the subharmonic perturbations could be provided by the induced field of the downstream (already paired) vortices, and the phase reference through a feed-back loop fixed to the trailing edge of the nozzle. Thus, a tentative picture of the interaction may be imagined in the following way (Figure 8). A pairing interaction occurring periodically at $x = x_p$ creates an induced field perturbing the vortices generated at $x = 0$ (see tagged vortices in Figure 8). The perturbed vortices interact with their neighboring one to form the larger vortices at $x = x_p$. For resonance, it is necessary that the

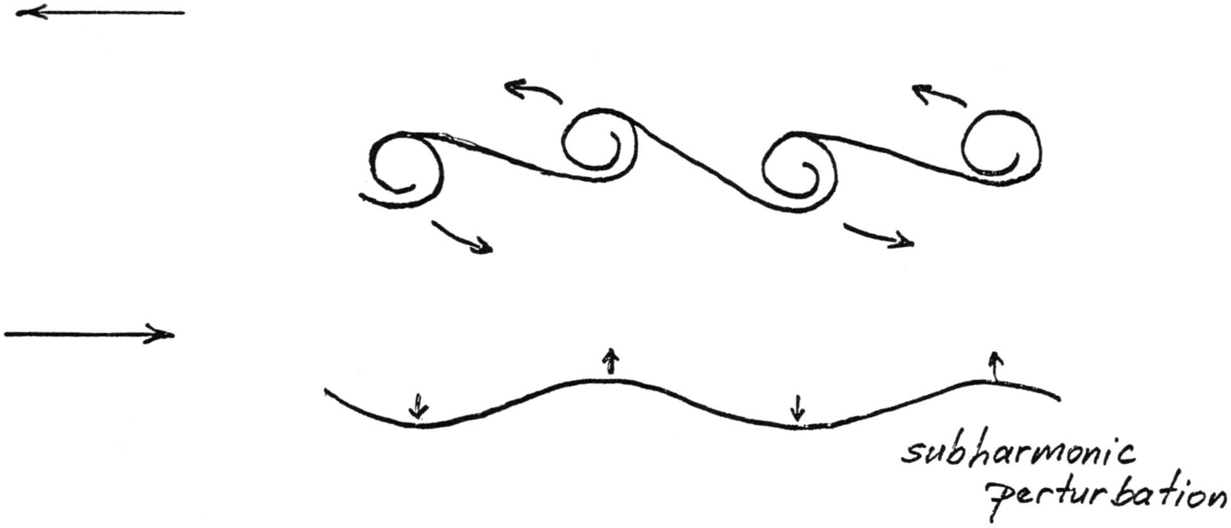

subharmonic perturbation

Figure 7 Subharmonic Interaction

period required for the incipient vortex to reach the pairing location plus the time for the disturbance signal generated by the interaction to feed back to the nozzle edge be an integer number times the resonance period. Accordingly,

$$\frac{x_p}{V_c} + \frac{x_p}{a} = \frac{N}{f_r}$$

or

$$\frac{fD}{V_j} \frac{x_p}{D} \left(\frac{V_j}{V_c} + \frac{V_j}{a}\right) = N$$

where V_c, a and V_j are the vortex convection, ambient sound and jet velocities respectively. This is a relation identical to that proposed by Ho and Nosseir for the resonance in the jet impingement problem.[3]

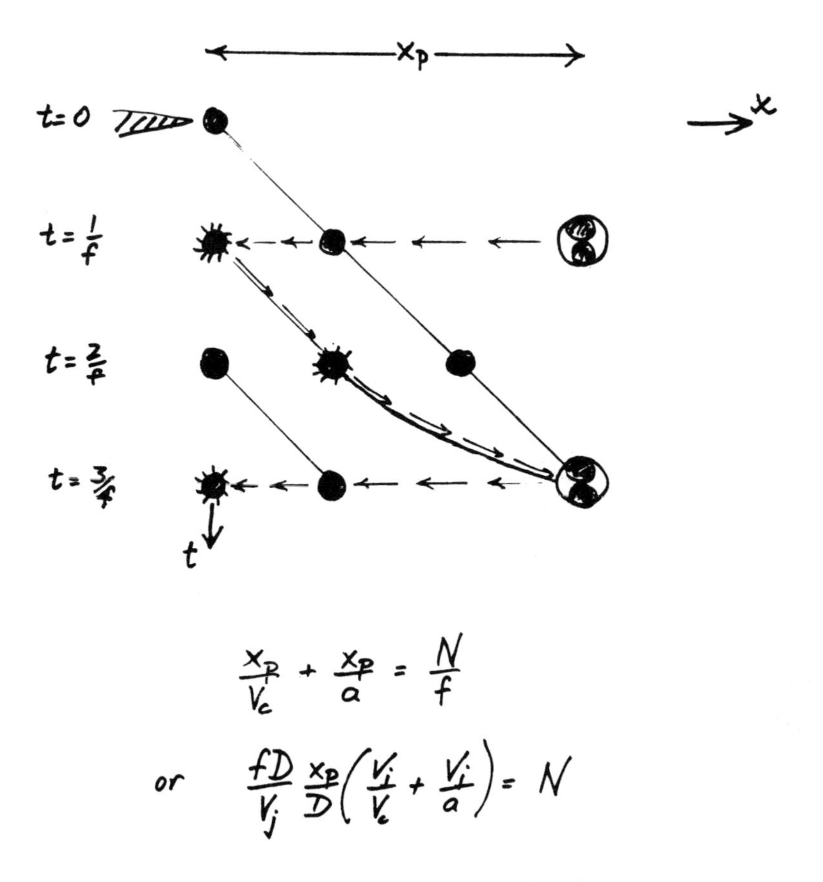

Figure 8 Subharmonic Feedback

It is not expected, of course, that in a free jet a phase lock or pure tone resonance would occur: changes in any link of the feed-back loop will cause phase fluctuations and a time variation in the magnitude of x_p. An experimental confirmation of the above mechanism would be greatly facilitated, if these fluctuations could be minimized. This, in fact, was done by Kibens[7] who introduced a small, axisymmetric, periodic perturbation onto the shear layer at the nozzle exit having a frequency equal to either the instability frequency of the layer or its first subharmonic. With this method he greatly reduced the relative random modulation amplitude of the shear layer oscillations imposed by external disturbances (free stream turbulence etc.). As a result, the pairing locations in the mixing layer were fixed and the mixing region of the jet became an organized, column of air with periodic oscillations. Figure 9 summarizes his results. The initial shear layer instability corresponds to $St = 3.54$ which is also the excitation frequency. It is seen that this frequency is detectable up to $x/D = 0.5$ at which station the first pairing process is terminated and only the first subharmonic can be observed. The second and third pairings are seen to occur at $x/D = 1.0$ and 2.5.

The straight line in the figure corresponds to the above equation with $N = 3$. It is seen that the predicted pairing locations are in reasonable agreement with the experiments giving encouragement for further inquiry along the same lines.

Conclusion

It is seen from the above example that the initial instability of a free shear layer plays an important role in the subsequent development of the flow. It is also seen that the results described were deduced using a deterministic point of view: the vortex structures generated by the unstable shear layer interact with each other through an induction mechanism. The random behavior is caused by outside and self induced disturbances that produce phase fluctuations in the interaction process. Finally, if the above description of the flow development is a realistic one, one must conclude that the turbulent transport process is actuated by a long range force field, the induced pressure field and therefore cannot be expressed in terms of a local gradient diffusion model.

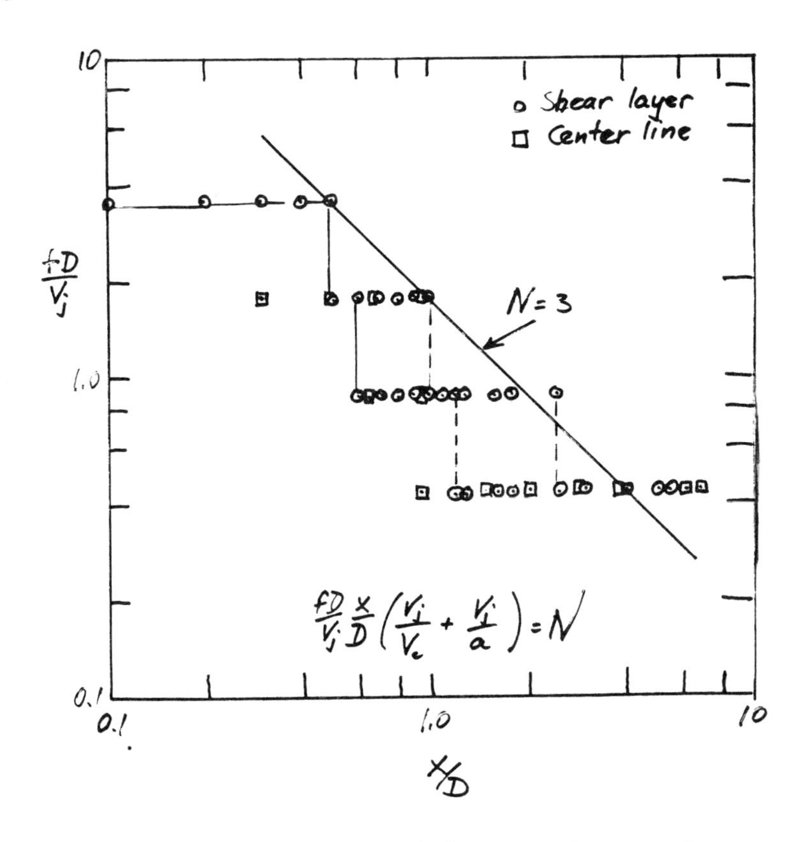

Figure 9 The forced jet experiments of Kibens (Ref. 7).

REFERENCES

1. Brown, G. L. and Roshko, A., On density effects and large structures in turbulent mixing layers, J. Fluid Mech., 64 (1974) 775-816.
2. Winant, C. D. and Browand, F. K., Vortex pairing: The mechanism of turbulent mixing layer growth at moderate Reynolds number, J. Fluid Mech., 63 (1974) 237-255.

3. Ho, C. M. and Nosseir, N. S., Dynamics of an impinging jet, Part 1: The feedback phenomenon (J. Fluid Mech., to be published).
4. Laufer, J. and Monkewitz, P., On turbulent jet flows: A new perspective, AIAA Paper 80-0962, Hartford, Conn. June, 1980.
5. Patnaik, P. C., Sherman, F. S. and Corcos, G. M., Vorticity concentration and the dynamics of unstable free shear layers, J. Fluid Mech. 73 (1976) 215-241.
6. Riley, J. J. and Metcalfe, R. W., Direct numerical simulation of a perturbed, turbulent mixing layer, AIAA Preprint 80-0274, 1980.
7. Kibens, V., Discrete noise spectrum generated by an acoustically excited jet, AIAA J., 18 (1980) 434-441.

Department of Aerospace Engineering
University of Southern California
Los Angeles, CA 90007

Instability and Transition in Pipes and Channels

J. T. Stuart

1. INTRODUCTION

It has been known since the time of Osbourne Reynolds in the latter part of the 19th century that laminar flow in a pipe or rectangular channel is frequently and naturally replaced by oscillatory motion, made visible by dye filaments, or by "slugs" of turbulence, interspersed with regions of calm or laminar flow, or by a more complete random turbulent motion. Theoretical developments followed, with a distinction to be drawn between classes of flows which are unstable (I) or stable (II) with respect to infinitesimal perturbations. We note prototype examples in turn of each of these categories.

I. Plane Poiseuille flow, driven by a pressure gradient, between two parallel planes.

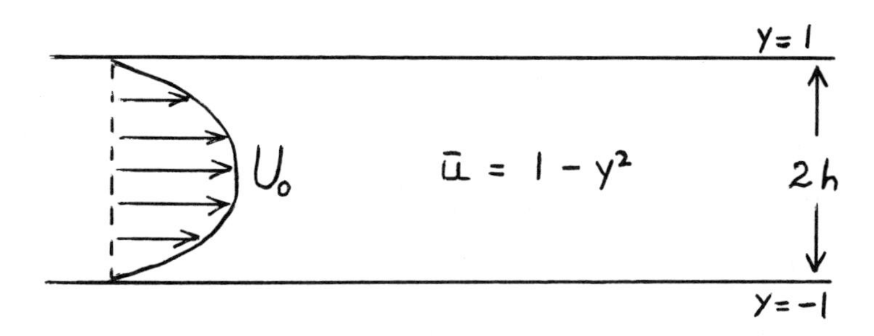

ISBN 0-12-493240-1

The Reynolds number, R, is $U_0 h/\nu$, where U_0 is the maximum velocity, 2h is the distance between the planes and ν is the kinematic viscosity. Transition from laminar flow to turbulence occurs experimentally in the range of R from 1000 to 2500 (Davies and White, Tillmann); but instability to infinitesimal disturbances occurs theoretically and experimentally for Reynolds numbers above a value of about 5800.

IIa. Plane Couette flow, driven by the motion of one of two parallel planes.

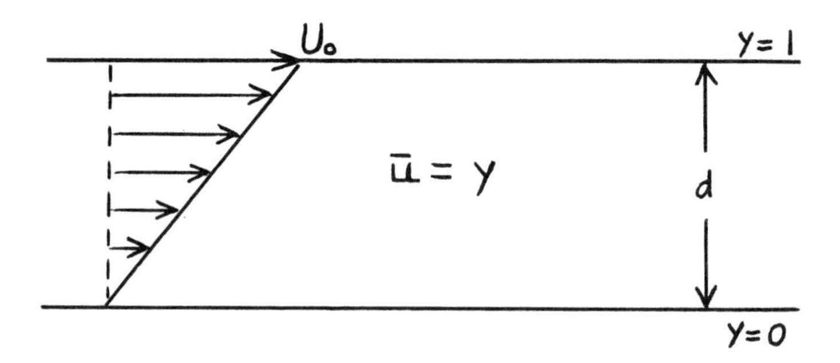

The Reynolds number, R, is $U_0 d/\nu$, where U_0 is the maximum velocity and d is the distance between the planes. Transition to turbulence occurs experimentally for Reynolds numbers in the range 1000 to 2000 (Taylor, Reichardt); it is generally accepted that instability to infinitesimal disturbances does not occur, though this has never been rigorously proven.

IIb. Hagen-Poiseuille flow, driven by a pressure gradient, in a circular pipe.

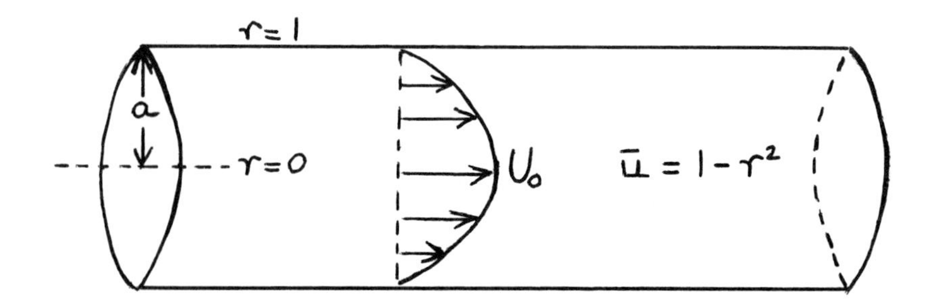

The Reynolds number, R, is $U_o a/\nu$, where U_o is the maximum velocity and a is the radius of the pipe. If the flow is kept very undisturbed, the flow in experiment can be kept laminar for values of R up to 50,000 (Ekman, Taylor); otherwise, if finite-amplitude disturbances are allowed, turbulence can occur for R as low as about 2000. As with plane Couette flow it is generally accepted that instability to infinitesimal disturbances does not occur, though this has never been rigorously proven.

The distinction between cases I and II requires them to be treated separately, and this procedure will be followed in this lecture.

2. HISTORICAL BACKGROUND: THEORY

(a) Reynolds (1880-1895) introduced the concept of transport of momentum by finite velocity fluctuations, which gives rise to the apparent stress now known as the Reynolds stress.

(b) Rayleigh (1880-1916) discussed waves in inviscid shear flows, introducing the notion of the presence of an inflection point in the velocity profile being relevant for instability. (It was later shown by Fjørtoft in 1950 and Høiland in 1953 that an additional necessary condition for instability is that the magnitude of the vorticity must be a maximum at the inflection point.)

(c) The Orr-Sommerfeld equation, which governs viscous vorticity waves in plane shear flows was enunciated in 1907.

(d) Prandtl and Tietjens (1925, 1935) gave an account of the physical origin of the instability mechanism, which involves the effect of a phase shift in the velocity fluctuations to produce a flow of energy from the basic flow to the fluctuations. The contribution of Tietjens was an incomplete calculation of instability, which did not produce a critical Reynolds number.

(e) Heisenberg (1925) in his doctoral thesis overcame severe mathematical problems, especially in association with the "critical (or Heisenberg-Tollmien) layer", to show that plane Poiseuille flow is unstable at large Reynolds numbers.

(f) Tollmien (1929, 1935) and Schlichting (1933 and

later) developed instability theory for the important case of boundary layers, including particularly the Blasius boundary layer on a flat plate. In particular, Tollmien (1929) was the first to obtain a critical Reynolds number for waves in a nearly-parallel shear flow, although this had been done by G.I. Taylor in 1923 for flow between rotating cylinders.

(g) Tollmien's predictions were substantially confirmed by the elegant and beautiful experiments of Schubauer and Skramstad during the years 1940 - 1943.

3. BASIC EQUATIONS AND BASIC FLOW

Let $\underset{\sim}{x}$ represent a position vector, t the time, $\underset{\sim}{u}$ the velocity, $\underset{\sim}{\omega}$ the vorticity, p the pressure, ρ the density and ν the kinematic viscosity. Then the governing equations are

$$\frac{\partial \underset{\sim}{u}}{\partial t} - \underset{\sim}{u} \wedge \underset{\sim}{\omega} = - \nabla\left(\frac{p}{\rho} + \frac{1}{2} \underset{\sim}{u}^2\right) - \nu \operatorname{curl} \underset{\sim}{\omega}, \tag{3.1}$$

$$\underset{\sim}{\omega} = \operatorname{curl} \underset{\sim}{u}, \tag{3.2}$$

$$\nabla \cdot \underset{\sim}{u} = 0. \tag{3.3}$$

The basic flow, whose stability is to be examined, is often a simple steady solution of these equations, independent of the coordinate in the direction of that basic flow. Thus, for plane Poiseuille flow, the velocity component parallel to the planes is $\bar{u}(y)$, where y is the coordinate normal to them; on the other hand, for Hagen-Poiseuille flow the velocity parallel to the pipe walls is $\bar{u}(r)$, where r is the radial coordinate. Sometimes the basic flow is time dependent, as in several papers of P. Hall, including that treating the plane Stokes' layer on an oscillating wall; and sometimes it is varying in the flow direction, as in papers of Bouthier, DiPrima and Stuart, Gaster, Nayfeh and Saric, and F.T. Smith.

4. LINEARIZED PERTURBATIONS IN PLANE POISEUILLE FLOW

It is known from the work of Squire (1933) that the minimum critical Reynolds number, above which instability can take place, occurs for two-dimensional, or plane, perturbations. The stream function is thus written as

$$\Psi = y - \frac{1}{3} y^3 + \psi(y) \exp\{i\alpha(x - ct)\}, \tag{4.1}$$

where $y = \pm 1$ represent the planes, x is the streamwise coordinate, the term independent of x and t represents the basic flow, and the exponential term represents the linearized wavy perturbation. There are two basic possibilities:

(i) α is real, c is complex, growth rate in time is αc_i;

(ii) α is complex, $\alpha c = \omega$ is real, growth rate in space is $-\alpha_i$.

The growth rates, if suitably small, are related by the Gaster transformation, which involves the group velocity of the waves.

The perturbation stream function ψ satisfies the Orr-Sommerfeld equation:

$$(1-y^2-c)(\psi'' - \alpha^2\psi) + 2y\psi + i(\alpha R)^{-1}(\psi^{IV} - 2\alpha^2\psi'' + \alpha^4\psi) = 0; \quad (4.2)$$

$$\psi = \psi' = 0, \; y = \pm 1. \quad (4.3)$$

The solution which gives instability, and which is believed to lead to a uniquely unstable mode, has ψ even in y and leads to the eigenvalue relation:

$$F(\alpha^2, c, \alpha R) = 0. \quad (4.4)$$

For instability, it is known that αR is of order 5800, so that ψ has Stokes' boundary layers on the walls, and Heisenberg-Tollmien critical layers neighbouring (complex) values of y, where $1 - y^2 = c$.

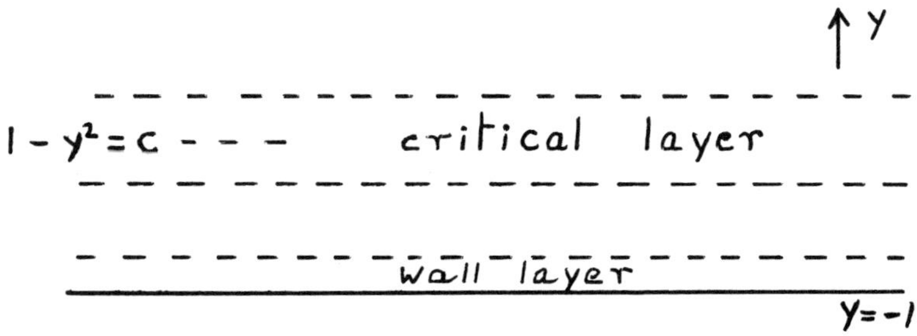

y

$1 - y^2 = c$ critical layer

wall layer

$y = -1$

Heisenberg obtained the asymptotic behaviour of F for $\alpha \to 0$ and $R \to \infty$, with $\alpha R \to \infty$ and established instability between the two asymptotes in the α, R plane (see Figure below); other significant contributions have been made by Lin, Meksyn, Thomas, Davey, Lakin, Ng and Reid, Eagles, de Villiers and F.T. Smith.

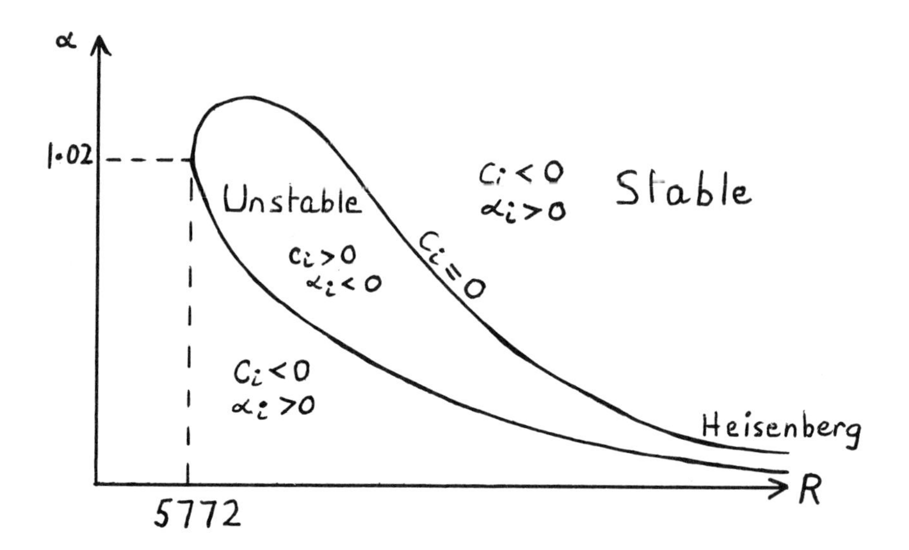

Earlier Prandtl had shown that energy is transferred from the basic flow to the wave motion through the action of the Reynolds stress. The figure below shows a typical distribution of Reynolds stress, plotted as a function of y, with a notable peak at the critical layer.

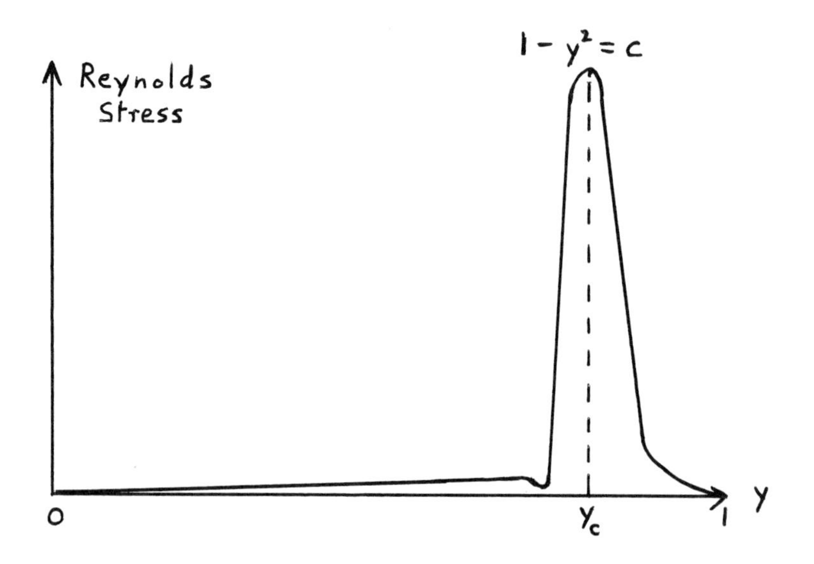

Finally, we note Squire's result that three-dimensional perturbations proportional to exp $\{i(\alpha x + \beta z - \alpha ct)\}$ lead to the eigenvalue relation

$$F(\alpha^2 + \beta^2, c, \alpha R) = 0 \tag{4.5}$$

where the function, F, is the same as in formula (4.4).

5. A CRUCIAL DIFFICULTY IN PLANE POISEUILLE FLOW

It is seen from the results illustrated in Section 4 that the minimum critical Reynolds number occurs for R = 5772 approximately. On the other hand, turbulence occurs at Reynolds numbers as low as 2500 or even 1000. Thus it is not possible to argue that turbulence would follow as a development from the linearized perturbations of the previous section, by raising the Reynolds number.

One possibility is that nonlinear effects are important. Heisenberg wrote down the relevant equations in 1925, but did not solve them. Important nonlinear processes involve the development of

(i) harmonics, exp $\{in\alpha(x - ct)\}$, n is an integer;

(ii) a change of the mean flow $\bar{u}(y)$ by the Reynolds stress.

A crucial question is this: can the nonlinear terms promote instability above some threshold amplitude, to be followed by turbulence at larger amplitudes? This idea has been the subject of much activity in the last thirty years, and is described in the next section.

6. NONLINEAR THEORETICAL DEVELOPMENTS IN PLANE POISEUILLE FLOW

Meksyn and Stuart (1951), followed by Grohne (1955), performed "mean-field" calculations, in which the generation of harmonics was excluded but in which the basic flow was allowed modification by the Reynolds stress. This led to the curve plotted below of a critical Reynolds number against the threshold amplitude, A_e, of the fundamental wavy perturbation. The lowest Reynolds number on this curve is about 2900, with a root-mean square velocity fluctuation of some 8%. Only the equilibrium state was calculated, but the work implies that there is instability above the threshold curve and stability below.

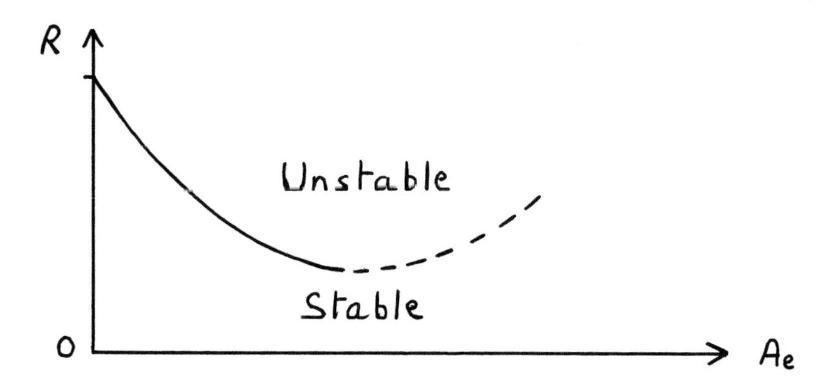

A rational theory, which includes the effect of harmonics, was constructed in 1960 by Stuart and Watson, and led to an "amplitude" equation which earlier (1944) had been adumbrated by Landau. The 1960 theory considers a perturbation from the basic flow of the form

$$A(t)\psi(y)\ \exp\{i\alpha(x - ct)\} + \ldots \tag{6.1}$$

By detailed calculation, it can be shown that, for small growth rates αc_i, the amplitude $A(t)$ must satisfy

$$\frac{dA}{dt} = \alpha c_i A + a_1 A|A|^2 + \text{higher order terms.} \tag{6.2}$$

Moreover, the square modulus satisfies

$$\frac{1}{2}\frac{d|A|^2}{dt} = \alpha c_i |A|^2 + a_{1r}|A|^4 + \text{higher order terms.} \tag{6.3}$$

First calculations of a_{1r} were made in 1967 by Pekeris and Shkoller and by Reynolds and Potter. Different conditions on the mean flow were maintained by these workers, a constant mean pressure gradient by Pekeris and Shkoller but a constant mass flux by Reynolds and Potter. The results, however, are qualitatively similar.

These results indicate that, in the region of the minimum critical Reynolds number, $R_c = 5772$, $\alpha_c = 1.02$, the number a_{1r} is positive. However, on the lower branch of the neutral curve, $a_{1r} < 0$ for values of R which are sufficiently large, as is indicated in the next figure. (There are also important three-dimensional effects on the lower branch.)

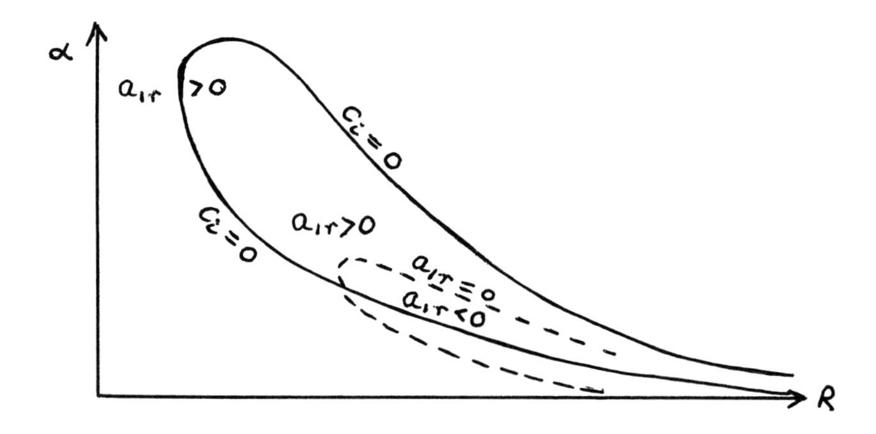

The most significant result is that $a_{1r} > 0$ near to the critical Reynolds number, since this implies that a threshold amplitude exists below the critical Reynolds number, where $\alpha c_i < 0$. The solution of (6.3) is

$$|A|^2 = \frac{K\alpha c_i \exp(2\alpha c_i t)}{1 - Ka_{1r} \exp(2\alpha c_i t)}, \quad K < 0 \tag{6.4}$$

The following schematic diagram indicates the threshold amplitude, with instability at still greater amplitudes.

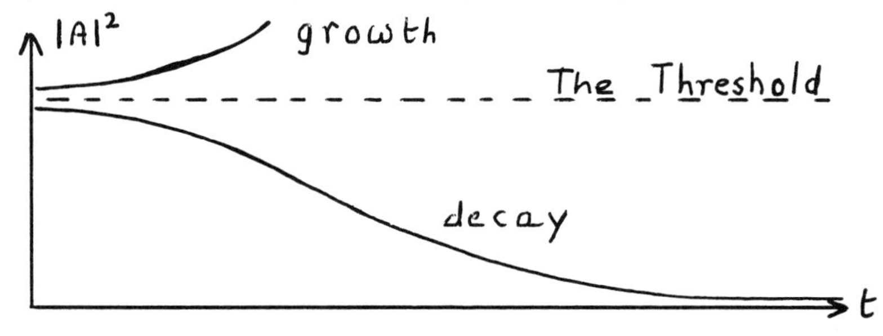

At Reynolds numbers much lower than the value 5772, it may be expected that a similar threshold phenomenon exists, and important work of Fasel, T. Herbert, and Orszag and Kells has shown this; this confirms that the results of Meksyn and Stuart are qualitatively correct and, indeed, give a critical Reynolds number which is fairly accurate.

7. EXPERIMENTS IN PLANE POISEUILLE FLOW

In 1975 Nishioka, Iida and Ichikaya published the results of an experimental investigation of the linear and nonlinear aspects of instability in plane Poiseuille flow. Their measurements were compared with detailed theoretical investigation by N. Itoh (1974), of the type already described in Sections 4 and 6. The channel had a cross section of aspect ratio 27:1, thus reducing the effects of the side walls. To generate waves propagating in the x direction, they used the classical "vibrating-ribbon" technique of Schubauer and Klebanoff.

The results may be summarized as follows:

(i) Quite good agreement was obtained for the neutral curve and unstable zone of linearized theory;

(ii) A calculation by Itoh of the behaviour of the root-mean square threshold velocity fluctuation, as a function of frequency at R = 5000, showed good agreement with experiment for a range of frequencies. This is indicated in the figure below.

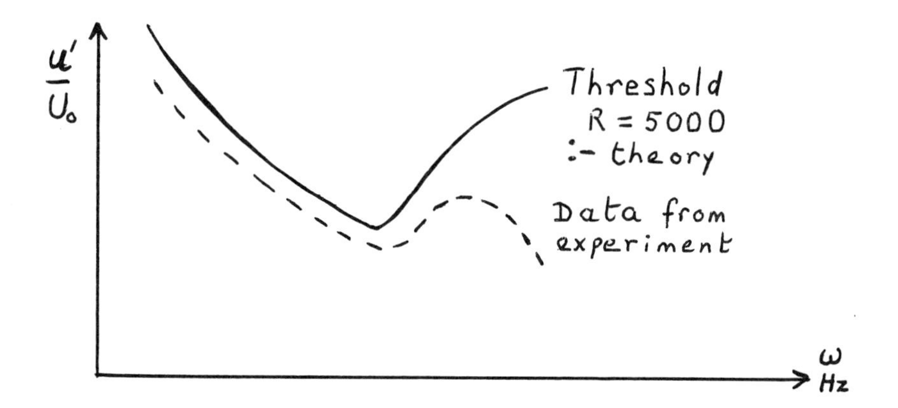

As a consequence of the comparison of theory and experiment, we may regard the threshold-amplitude phenomenon as an established fact, at least in plane Poiseuille flow.

8. NONLINEAR CRITICAL LAYERS

In the Orr-Sommerfeld equation (4.3) there is a critical layer where the essential balance is between the terms $(1 - y^2 - c)\psi''$ and $i(\alpha R)^{-1}\psi^{IV}$, and the viscous critical-layer thickness is therefore proportional to $(\alpha R)^{-1/3}$. In 1968, Benney and Bergeron and, separately, Russ H. Davis, showed that a different balance of terms is possible to eliminate the logarithmic singularity associated with $(1 - y^2 - c)\psi''$. Non-linear terms, of order $\psi'\psi''$, have been omitted from (4.3); if a balance is made with those terms, the critical layer thickness is proportional to $\varepsilon^{1/2}$, where ε is the local amplitude of ψ. Viscous terms become necessary only at higher order for this nonlinear critical layer.

The ratio of the viscous and nonlinear critical-layer thicknesses is $R_L^{1/2}$, where $R_L = \varepsilon R^{2/3}$ is a local Reynolds number for the oscillation in the neighbourhood of the critical layer. If R_L is small, the viscous theory applies (equation 4.3); but if R_L is large, the critical layer is non-linear. Further developments of these ideas have been made by Kelly, Maslowe, and Brown and Stewartson.

More recently, F.T. Smith has developed a nonlinear theory of instability in the neighbourhood of the upper branch of the neutral stability curve. His work shows that the nonlinear critical layer arises quite naturally in an amplitude series.

9. HAGEN-POISEUILLE FLOW IN A CIRCULAR PIPE

We recognise this type of flow as a prototype of flows which are stable according to linearized perturbation theory. One important feature, which has been the subject of attention from Coles, Wygnanski and others, is the turbulent slug, which moves with some speed of propagation and has at front and rear rather sharp interfaces separating contorted, turbulent vorticity within from "calm", or laminar, flow outside. No satisfactory theoretical explanation of such slugs has been given. However, later in this lecture we shall discuss a possible theoretical basis for slowly-varying slugs of vorticity of long wavelength (we emphasize that is not a turbulent slug).

As a preliminary we need to consider the linearized instability theory, which has been discussed over the years by Sexl (1927), Davey and Drazin, Garg and Rouleau, Gill, Lessen, and Grosch and Salwen. If r, θ, z are cylindrical polar coordinates, perturbations may be considered of the form

$$\exp\{i(\alpha z + n\theta - \omega t)\}, \tag{9.1}$$

where n is an integer, to preserve a single-valued solution. Exponential growth would occur for $\omega_i > 0$, and decay for $\omega_i < 0$.

A large amount of detail is known of these instability modes as to their radial structure, "wall" modes and "centre" modes being one clear distinction. The least stable modes are known to have $n = 1$, but it is believed nevertheless that all linearized modes are stable, although this has never been proved rigorously. This is a serious problem, not unlike the difficulty in plane Poiseuille flow as was exposed in Section 5.

There are several ways in which theoretical attempts have been made to pinpoint an instability process in Hagen-Poiseuille flow:

(a) Tatsumi (1952) showed that the basic developing flow in the entry length of a pipe is unstable to linearized perturbations at or above a Reynolds number of order 10^4. This may be relevant for some experiments but, if the instability were to persist into the fully developed flow region, some nonlinear or other influence would be needed.

(b) Davey has considered a problem with an imperfection, by studying the instability of flow in a slightly elliptical pipe of eccentricity ε. By expansion in terms of the latter parameter, the growth rate ω_i is found to be of the form

$$\omega_i = -\sigma_o + \varepsilon^2\sigma_1, \tag{9.2}$$

where $\sigma_o > 0$, $\sigma_1 > 0$.

Without the calculation of higher-order terms, one cannot be sure that instability has been achieved, since higher-order terms might reverse the ε^2 contribution to the amplification rate. It is, however, established by Davey's work that very small values of ε produce a term $\varepsilon^2\sigma_1$ comparable with σ_o.

(c) Davey and Nguyen, Itoh and Davey have developed a theory like that of Section 6, and I believe that their work shows that the nonlinear terms in (6.3) may be destabilizing in this Hagen-Poiseuille case for sufficiently small amplitudes. There is, however, no small parameter, so that one cannot estimate the influence of higher-order terms; in particular, the existence of a threshold amplitude has not been established.

10. G.I. TAYLOR'S LONG-WAVELENGTH THEORY

Later we wish to discuss the transport of vorticity fluctuations of long wavelength by Hagen-Poiseuille flow in a pipe. The problem is somewhat analogous to that of the diffusion and mass transport of a passive scalar, such as a contaminant, which was discussed by Taylor in 1954. He showed that, because of an essential balance between radial diffusion and longitudinal convection, a contaminant in Hagen-Poiseuille flow satisfies the one-dimensional diffusion equation as though it were diffusing along the pipe, but with an effective velocity-dependent coefficient of diffusion. As a preliminary to our vorticity investigation we discuss Taylor's theory by means of a multiple-scaling argument. For related work see the thesis of R.J. Jefferson (1978); different methods have been described by Chatwin and others. If the pipe has non-dimensional radius 1, a scalar contaminant, C, satisfies the equation

$$\frac{\partial C}{\partial T} + (1 - r^2)\frac{\partial C}{\partial z} = \lambda_D\left(\frac{\partial^2 C}{\partial r^2} + \frac{1}{r}\frac{\partial C}{\partial r} + \frac{\partial^2 C}{\partial z^2}\right), \tag{10.1}$$

where

$$\lambda_D = D/U_o a, \tag{10.2}$$

a being the radius of the pipe, U_o the maximum velocity and D the diffusivity. Radial and axial lengths, r and z, are scaled on a, and time T on a/U_o. There is a boundary condition

$$\frac{\partial C}{\partial r} = 0, \quad r = 1 \tag{10.3}$$

and a condition that C is regular and that the flux is given in an appropriate moving reference frame.

There is a simple eigensolution of (10.1) and (10.3), namely

$$C = \overline{C} = \text{constant} \tag{10.4}$$

We now introduce long time (δ^{-2}) and length (δ^{-1}) scales

$$\tau = \delta^2 T, \quad \zeta = \delta(z - \mathcal{C} T), \tag{10.5}$$

where $\mathcal{C}$ is a propagation speed. We now let C depend on r, τ and ζ and assume

$$C = \bar{C}(r,\tau,\zeta) + \delta C_1(r,\tau,\zeta) + \delta^2 C_2(r,\tau,\zeta) + \ldots \tag{10.6}$$

On substituting (10.5) and (10.6) in (10.1) we find

$$O(\delta^0) \;:\; 0 = \frac{\partial^2 \bar{C}}{\partial r^2} + \frac{1}{r}\frac{\partial \bar{C}}{\partial r} \Rightarrow \bar{C} = \bar{C}\,(\tau,\zeta) \;: \tag{10.7}$$

$$O(\delta) \;:\; (1 - r^2 - \mathcal{C})\frac{\partial \bar{C}}{\partial \zeta} = \frac{\lambda_D}{r}\frac{\partial}{\partial r}\left(r \frac{\partial C_1}{\partial r}\right) \tag{10.8}$$

$$\frac{\partial C_1}{\partial r} = 0, \quad r = 1. \tag{10.9}$$

Equation (10.8) subject to (10.9) has a solution only if the left hand side is orthogonal to the solution adjoint to (10.7), which is $\bar{C}$ because of self-adjointness. This gives

$$\mathcal{C} = \tfrac{1}{2}. \tag{10.10}$$

If we insist on the mean concentration over a section being defined as $\bar{C}$, then we have

$$\int_0^1 r\, C_1\, dr = 0, \tag{10.11}$$

and

$$C_1 = (8\lambda_D)^{-1}\left[-\tfrac{1}{3} + r^2 - \tfrac{1}{2} r^4\right]\frac{\partial \bar{C}}{\partial \zeta} \tag{10.12}$$

$O(\delta^2)$: An equation for C_2 is obtained in which, by use of (10.12), $\partial\bar{C}/\partial\tau$ and $\partial^2\bar{C}/\partial\zeta^2$ appear. A condition of solubility on this equation for C_2 yields

$$\frac{\partial \bar{C}}{\partial \tau} - \left(\lambda_D + \frac{1}{192\lambda_D}\right)\frac{\partial^2 \bar{C}}{\partial \zeta^2} = 0. \tag{10.13}$$

With dimensional variables t and $Z = az - \frac{1}{2}U_0 t$, this becomes

$$\frac{\partial \bar{C}}{\partial t} - \left(D + \frac{U_0^2 a^2}{192 D}\right)\frac{\partial^2 \bar{C}}{\partial Z^2} = 0. \tag{10.14}$$

Equation (10.14) is the diffusion equation, for one-dimensional diffusion <u>along</u> the pipe. In the diffusion coefficient, D represents molecular diffusion, whereas the <u>Taylor</u>

coefficient $U_o^2a^2/192D$ represents an effective diffusion brought about by an interplay of axial convection and radial diffusion. The Taylor effect is far more important than molecular diffusion if $U_oa/D >> 14$.

We note that (10.14) can be used to describe the evolution of a slug of contaminant. Our object now is to use similar ideas, including multiple scaling, to obtain an equation for the evolution of a slug of vorticity.

11. THE EVOLUTION OF VORTICITY OF LONG WAVELENGTH

Instead of treating convection and diffusion of the passive scalar C, we consider the Navier-Stokes equations and linearize about Hagen-Poiseuille axial flow. If the only perturbation to that basic flow is a swirl about the axis, the circulation Ω satisfies an equation similar to (10.11):

$$R^{-1}\frac{\partial\Omega}{\partial T} + \delta(1 - r^2)\,\frac{\partial\Omega}{\partial z} = R^{-1}\left(\frac{\partial^2\Omega}{\partial r^2} - \frac{1}{r}\,\frac{\partial\Omega}{\partial r} + \frac{\partial^2\Omega}{\partial z^2}\right). \qquad (11.1)$$

Here T is scaled on a^2/ν, and r and z on a, the radius of the pipe; moreover

$$R = U_oa/\nu. \qquad (11.2)$$

The boundary conditions are

$$\Omega = 0, \ r = 1, \qquad (11.3)$$

together with regularity. We return later to the matter of initial conditions.

An eigensolution of (11.1) and (11.2) is

$$\Omega = rJ_1(\lambda_jr)\,\exp\,(-\lambda_j^2T); \quad J_1(\lambda_j) = 0; \qquad j = 1,2,3,\ldots \qquad (11.4)$$

In contrast to (10.4), we note that this is r and T dependent.

We now introduce long time (δ^{-2}) and length (δ^{-1}) scales by (10.5), but rewrite the propagation speed as

$$\ell = cR. \qquad (11.5)$$

The governing equation (11.1) then becomes

$$\frac{\partial\Omega}{\partial T} + \delta^2\frac{\partial\Omega}{\partial\tau} + \delta R(1-r^2-c)\,\frac{\partial\Omega}{\partial\zeta} = \frac{\partial^2\Omega}{\partial r^2} - \frac{1}{r}\,\frac{\partial\Omega}{\partial r} + \delta^2\,\frac{\partial\Omega}{\partial\zeta} \qquad (11.6)$$

We now expand Ω as

$$\Omega = S(\tau,\zeta)\ r\ J_1(\lambda_j r)\ \exp(-\lambda_j^2 T) + \delta\Omega_1(T,r;\tau,\zeta) + \delta^2\Omega_2\ (T,r;\tau,\zeta) + \ldots, \quad (11.7)$$

and the results are as follows:

$O(\delta^o)$: The function S, which multiplies the eigenfunction (11.4), depends on τ and ζ only:

$O(\delta)$: An integrability condition on the equation for Ω_1 requires that

$$c = \frac{2}{3}, \quad (11.8)$$

which is not the mean flow speed of Taylor's case;

$O(\delta^2)$: An integrability condition on the equation for Ω_2 requires that

$$\frac{\partial S}{\partial \tau} - K\frac{\partial^2 S}{\partial \zeta^2} = 0, \quad (11.9)$$

$$K = 1 + \frac{R^2}{15\lambda_j^4}\left(8 - \frac{1}{3}\lambda_j^2\right). \quad (11.10)$$

The expansion can be taken to higher order, but we rest at this stage.

In the table below the first four values of λ_j are given:

j	1	2	3	4
λ_j	3.83171	7.01559	10.17347	13.32369
$R_c(j)$	-	66	78	96

For j = 1, K is positive for all values of the Reynolds number; but, for j = 2 and for greater values, K changes sign from positive to negative values at $R = R_c(j)$, whose values are given in the table. Thus:

$$K < 0, \quad R > R_c(j), \quad j = 2, 3, 4, \ldots \quad (11.11)$$

We return later to the meaning of this negative diffusion coefficient.

It is possible to do calculations similar to that given above for (i) non-swirling axisymmetric modes and (ii) for general non-axisymmetric modes. Details of this work will be recorded elsewhere, but it is noted here that negative diffusion can occur.

12. POSITIVE OR NEGATIVE EFFECTIVE DIFFUSION

Problems with positive diffusion ($K > 0$) are, of course, common in Mathematics and in the Physical Sciences. If, at $\tau = 0$, S is specified as a function of ζ, it spreads out and decays in magnitude as τ increases.

The situation with negative diffusion is the reverse of this. Consider the following problem:

$$\frac{\partial S}{\partial \tau} + (-K)\frac{\partial^2 S}{\partial \zeta^2} = 0, \quad -K > 0, \tag{12.1}$$

$$\tau = 0, \; S = a\exp(-\sigma^2\zeta^2), \tag{12.2}$$

$$|\zeta| \to \infty, \quad S \to 0. \tag{12.3}$$

The solution is

$$S = a(1 + 4K\tau\sigma^2)^{-1/2} \exp\{-\sigma^2\zeta^2(1 + 4K\tau\sigma^2)^{-1}\}. \tag{12.4}$$

As $\tau \to (-4K\sigma^2)^{-1}$, with ζ fixed and not equal to zero, S tends to zero. If, on the other hand, $\tau \to (-4K\sigma^2)^{-1}$, with $\zeta^2(1 + 4K\tau\sigma^2)^{-1}$ held fixed, then S tends to infinity. Thus the function S concentrates and amplifies as $\tau \to (-4K\sigma^2)^{-1}$.

It is instructive to consider further the occurrence of negative diffusion in this problem. If we consider a swirling perturbation of the form

$$\exp\,(pT + i\alpha z), \tag{12.5}$$

an expansion for small α (long wavelength) gives

$$p = p_o + \alpha p_1 + \alpha^2 p_2 + \ldots, \tag{12.6}$$

so that

$$p_o = -\lambda_j^2, \; p_1 = -\frac{2}{3}\, i\, R, \; p_2 = -K, \ldots\; . \tag{12.7}$$

It follows that (12.5) has the form

$$\exp\{-(\lambda_j^2 + K\alpha^2)T + i\alpha(z - \tfrac{2}{3}RT)\} \tag{12.8}$$

as an approximation for small α. We note that, if K is negative, waves of small α are less damped than those with $\alpha = 0$. Thus, negative diffusion here represents a flow of energy towards modes with less damping.

Referring back to (11.7), we note that the perturbation of swirl is decaying due to the exponential factor. But the negative diffusivity, when it occurs, causes a concentration and focussing of energy within the decaying slug. Thus the negative diffusion acts to reverse the natural decay. No non-linearity has been incorporated in this calculation. Moreover, the slug considered here is slowly-varying, and does not have the sharp interfaces which are characteristic of turbulent slugs. Nevertheless, we believe that the methods of analysis presented here, and the results, are of some interest.

This lecture was prepared and written as part of the Fluid Mechanics Program of the National Science Foundation through Grant (ME 78-22127 AOI, and of the Fluid Dynamics Program of the Office of Naval Research, at the Division of Engineering, Brown University. The writer is grateful to Professor Joseph T.C. Liu for his hospitality and for many discussions, and wishes also to thank Professors D.J. Benney, H.P. Greenspan and L.N. Howard, all of M.I.T., for their useful comments, especially on Sections 11, 12. Finally, he thanks the U.S. Army Research Office in London, together with Professors John Nohel and Richard Meyer, for their financial support for travel.

Mathematics Department
Imperial College
London SW7 2BZ
England

On Transition to Turbulence in Boundary Layers

M. Gaster

1. INTRODUCTION

My research is concerned with the transition processes that occur in boundary layer flows such as those that arise on the surfaces of smooth aerofoils. For simplicity, in both experiment and analysis, this is idealised to the problem of transition in the boundary layer of a flat plate at zero incidence to the flow. I believe that there has been a considerable amount of muddled thinking about some of the fundamental issues arising in this problem, and I want to draw attention to these. In particular, I shall discuss the type of stability problem that is posed - is it a spatial or temporal process? - or indeed is it a problem of stability at all? In fact the processes that lead to turbulence, at least in the boundary layer case, are not of stability or instability, but involve the evolution of forced oscillations. There are, of course, other flows that can properly be considered as posing a question of stability. Typically this occurs in a closed system, such as the Taylor-Couette flow that exists between rotating cylinders.

Let us first turn to this case of the Taylor flow. At low rotational speeds a steady base flow will be set-up without any periodic structure either along the length of the cell, or with respect to time. When this flow is perturbed by a small disturbance the motion returns to its base state -

ISBN 0-12-493240-1

it is therefore considered to be stable to small disturbances. This type of stability can be studied by examining solutions of the perturbed equations of the motion. The most general solution consists of functions involving spatial as well as temporal dependence. Any arbitrary initial disturbance can be decomposed into the various eigenmodes of the system, and if the imaginary part of the time dependent term of one of these modes is found to be positive the flow is deemed unstable. In fact it is now known that the first unstable mode, (the mode that appears when the critical rotational speed is marginally exceeded), is steady with respect to time, but has a periodic structure, of wavenumber k, along the cell length [1]. It turns out that there is a critical Taylor number, T_c, below which all small disturbances decay. At T_c, there is one particular value of the modal wavenumber, k_c, which is neutrally stable. Above this critical Taylor number there will be a band of unstable wavenumbers about k_c that exhibit exponential growth.

In practice when the flow is set up there is bound to be some residual motion, and these small disturbances act as seeds for the instability. Since we are concerned with exponential time growth even the very smallest disturbance will eventually become evident if the critical Taylor number is exceeded - this constitutes a true stability problem. Of course larger disturbance may well behave differently, but again one can ask the question - is the flow stable? - will it return to the base flow or not? If unstable the disturbance will presumably grow and another flow will result. Non-linear studies have shown that for some increase in Taylor number above the critical value steady finite amplitude cellular structures exist, with spanwise wavenumber close to the first linear mode. Further increase in Taylor number gives rise to wavy perturbations of these structures, and as the Taylor number is increased again more and more modes are possible. Whether the existence of a number of these non-linear modes interacting in some way can explain the onset of the random motion that we associate with turbulence remains an open question [2]. Nevertheless, the evolution of disturbances is clearly one of temporal instability, and it is natural to attempt to apply these ideas to the case of the boundary layer flow over a flat plate.

If a plate is put into a low-turbulence wind tunnel and the velocity fluctuations that arise in the boundary layer are examined with a the aid of hot-wire anemometer system, we can identify a number of different zones. Towards the front of the plate very little activity is seen at all and the fluctuations are really quite small. As the hot-wire is traversed downstream we first of all pick up small fluctuations, roughly periodic in character but fairly broad-band and modulated. Further downstream the amplitudes of these fluctuations get larger and at the same time the signal becomes more periodic. This narrow-band signal then begins to distort and the sinusoidal character of the signal occasionally breaks into bursts of high frequency oscillations. Very far downstream the signal becomes even less regular and contains more and more of these turbulent spots which spread and eventually coalesce to form a fully developed random motion. This suggests that, in some sense, we have a stability problem. Indeed, over roughly the last hundred years or so, much effort has been given to the study of the "stability" of parallel flows [3,4,5&6]. The boundary layer is not a strictly a parallel flow, and the linearized perturbation equations are partial differential equations that do not separate. Nevertheless, because the boundary layer grows relatively slowly it is often treated as if it were locally parallel, and this parallel "boundary layer" has indeed been treated precisely as a stability problem.

2. PARALLEL FLOWS

Consider the case of a base flow U(y) perturbed by some stream function $\psi(y,x,t)$. Following the stability concept of the Taylor problem one can ask whether, for sufficiently small amplitudes, there are solutions that grow or decay in time. In the case of the parallel base flow the linearized perturbation equations do separate and the Fourier transform method of selecting normal modes provides the general solution in the form

$$\psi(y,x,t) = \phi(y,\alpha,\omega)\exp i\{\alpha x - \omega t\} \tag{1}$$

where α is the wavenumber and ω the frequency. We are generally interested in those modes that have mainly real cigenvalues and then disturbances arise in the form of travelling waves. For a given wavenumber is there a value of α such that the imaginary part of ω is negative? A negative value of ω_i implies that the motion will amplify with respect to time in an exponential manner. The normal modes can be found as eigenvalues of the pertubation equations, just as for Taylor flow. Solutions of the Orr-Sommerfeld equation that are compatible with the four boundary conditions can readily be computed to provide a dispersion relation of the form

$$F(\alpha,\omega) = 0 \tag{2}$$

It has been established that in the case of a flat plate profile there are indeed critical values of the Reynolds number below which ω_i can be negative. This implies some form of temporal instability, and suggests that there might well be some dramatic change in the boundary layer flow where the Reynolds number exceeds the critical value. However, if a hot-wire is traversed downstream through the critical Reynolds number position, nothing significant occurs. It still seems to come as something of a surprise that we do not see anything spectacular. Indeed hardly any disturbances in this region are seen at all, either upstream or downstream of x_{crit}. Only if we go much further downstream do we begin to see signals containing any significant periodic character. The frequencies that appear in these oscillations lie fairly close to those of the eigenvalues of the Orr-Sommerfeld equation. We are, in fact, looking at the result of forcing by extraneous factors arising from imperfections in the experiment - from sound waves, from turbulence in the free-stream and, possibly, from various mechanical vibrations of the plate. Once generated these waves propagate downstream; some may well amplify, but I contend this is not stability in the sense of the closed Taylor problem.

In order to study these waves properly Schubauer and Skramstad [7] found it convenient to reduce the extraneous background random disturbances to a minimum and then to introduce controlled waves through a vibrating ribbon wavemaker. Indeed a very high proportion of experimental studies of boundary layer transition rely on wavemakers of some sort or another. The behaviour of the waves excited by a wavemaker can be calculated from the linearized perturbation equations of motion. Unfortunately these equations do not separate into ordinary differential form, and approximations have to be made. Neverthless, it is useful to consider first this hypothetical model of a "parallel boundary layer". The progress of a wavetrain that grows as it propagates downstream can also be determined from the characteristic function that has been used to define temporal modes, but interpreted in a slightly different way. These waves have a constant real frequency, ω , and they grow as they propagate. This can best be described by the so-called "spatial" modes defined in terms of the complex wavenumber [8],

$$\alpha = \alpha_r + i\alpha_i \qquad (3)$$

For small amplification rates values of α_i, defining the spatial growth, can be linked to the previously studied temporal growth ω_i through the group velocity [9]. This link is not a co-ordinate transformation, and it should be treated just as a convenient way of computing one type of growth in terms of the other.

3. WAVE PACKETS IN A PARALLEL FLOW

For other types of excitation we find that the structure of the solution is more closely modelled by modes with both complex wavenumber and complex frequency. This behaviour appears naturally in the following example of a wave packet. Consider the result of a two-dimensional wavemaker giving an isolated pulse in time at x_0 [10]. The motion will then be composed of a sum or integral of all possible waves, weighted in some way. If we assume that this weighting is reasonably flat, i.e. that all modes are equally excited, and that only the dominant mode is capable of amplification, then the result is given by

$$\Psi(y,x,t) = \int \phi(y,\alpha,\omega) \exp i\{\alpha x - \omega t\}\, d\omega . \qquad (4)$$

Writing (4) as

$$\Psi(y,x,t) = \int \phi(y,\alpha,\omega) \exp i\{Q(\omega)\, t\}\, d\omega,$$

The leading term of the asymptotic expansion as $t \to \infty$ is given by

$$\Psi(y,x,t) \to \sqrt{\frac{2\pi}{Q''(\omega^*)}} \cdot \exp i\{Q(\omega^*)t\} \times \left[1 + O\left(\frac{1}{t}\right) + \cdots\right] \qquad (5)$$

where the starred variables are associated with the value at the saddle-point of Q,

$$\frac{\partial Q}{\partial \omega}(\omega^*) = 0 \quad \text{or} \quad \frac{\partial \alpha}{\partial \omega}(\omega^*) = \frac{t}{x} \qquad (6)$$

Since we are interested in solutions for real values of x and t the quantity $\partial\alpha/\partial\omega$ is also real and may be identified with the reciprocal of the group velocity. Thus along any given ray there is a value of ω^* that satisfies equation (6). The solution, which is then given by equation (5), can be traced out as a wave packet in the x,t plane as shown in figure 1. In general the eigenfunction $\phi(y,\alpha(\omega^*),\omega^*)$ will be a solution of the Orr-Sommerfeld equation for complex values α and ω - it will be neither a purely spatial nor a purely temporal mode.

For the types of velocity profile found in attached boundary layers it turns out that both the leading and trailing edges of the wave packet travel downstream. The leading edge moves faster than the trailing edge so that the disturbed region of the flow expands as the packet propagates downstream. An impulsive excitation at a point will generate a patch of waves that propagates away from the source in the downstream direction. Waves within the packet may increase in amplitude as they propagate (as seen by an observer moving with the packet), but at any finite location downstream from the source point the motion will eventually decay - thus indicating stability. Such a behaviour pattern is illustrated on figure 1. However, there are flows containing wake or separated shear layer profiles, that do support solutions where the leading and trailing edges propagate in opposite directions. In these circumstances a pulse initiated at one station will generate a spreading patch of waves, but one that contains disturbances that grow exponentially near the source - in this case the motion is truly unstable (see figure 2). The disturbance, once excited, will grow with respect to time, and at some stage non-linear terms in the equations of motion will become important and a new non-linear motion must result. The linear disturbance must inevitably give way to a non-linear one. In such a flow one would expect a non-linear mode to appear quite naturally above the critical Reynolds number just as in the Taylor problem, where it is found that for a small increase above the critical Taylor number the new motion is a weak modification of the linear solution at the critical point. In the case of a wake, say, it seems quite likely that the linearly unstable motion will also be replaced by a neutral (constant amplitude) non-linear motion that has a similar structure to that of the linear mode. Indeed perhaps this is part of the explanation for the mechanism of frequency selection in the Karman vortex street that forms in the separated wake behind a body. The frequency of the most unstable linear eigenvalue of the wake profile is, in fact, quite close to that of the vortex street.

4. WAVE PACKETS IN A GROWING BOUNDARY LAYER

The boundary layer is not a parallel flow and the linearized equations describing the perturbation are partial differential equations that do not separate [11&12]. We seek solutions for the time periodic excitations $\exp i\omega t$ at x_0. The complete linearized equations contain small terms that prevent separation into ordinary differential form - leading to the "locally parallel flow approximation", that has already been discussed. In fact it is possible to organize the terms in the full equation so that an iterative process is set up to provide a successive sequence of corrections.

The solution of this scheme leads to a series definition of the streamfunction in inverse powers of the Reynolds number, the leading term being the locally parallel flow approximation modified by some scaling term A(x). The full solution appears as

$$\psi(y,x,t) = A(x,\omega)\, e^{i\left\{\int_{x_0}^{x}\alpha(x',\omega)dx' - \omega t\right\}} \left[\phi_0(y,x,\omega) + O\left(\tfrac{1}{t}\right) + \cdots\right] \tag{7}$$

where A(x) is defined by a first order linear differential equation.

A pulsed excitation of the growing boundary layer will therefore result in a wave packet defined by:

$$\psi(y,x,t) = \int A(x,\omega)\, e^{i\left\{\int_{x_0}^{x}\alpha(x',\omega)dx' - \omega t\right\}} \left[\phi_0 + O\left(\tfrac{1}{t}\right) + \cdots\right] d\omega\,. \tag{8}$$

Again writing the leading term in the form

$$\psi = \int A(x,\omega)_x\, \phi_0(y,x,\omega) \exp i\left\{Q(x,\omega)t\right\} d\omega, \tag{9}$$

where now Q is

$$Q(x,\omega) = \int_{x_0}^{x} \frac{\alpha(x',\omega)\,dx'}{t} - \omega,$$

we can find the leading term of the steepest descent asympote as before

$$\psi \to \sqrt{\frac{2\pi}{Q''(\omega^*)}}\, e^{iQ(\omega^*)t} \left[\phi_0(y,\alpha(\omega^*),\omega^*) + O\left(\tfrac{1}{t}\right) + \cdots\right], \tag{10}$$

where

$$\int_{x_0}^{x} \frac{\partial\alpha}{\partial\omega}(x',\omega^*)\,dx' = t\,. \tag{11}$$

For a given x and t we can find ω^* and hence the solution in that neighbourhood. But in this case there are no real <u>rays</u> defining trajectories of the various wave groups; the ray-like behaviour only exists in complex time. Although this concept of complex x or complex t may be useful, the result as given by (10) & (11) is more appropriately treated as an approximation to the integral defining the flow, and then the ray properties do not specifically enter into the discussion. These equations were solved numerically using the series dispersion relation discussed in reference [13]. For each streamwise station x an iterative procedure was set up to find ω^* as a function of t such that (11) was satisfied. It turned out that the mapping of ω^* onto the x, t plane was one-to-one, as shown on figure 3. As for the earlier parallel flow example the structure of the solution at a particular point and time can again be described in terms of eigenfunctions of the Orr-Sommerfeld equation for specific values of the wavenumber and frequency (starred values). At the centre of a packet ω_i^* is zero and the modal structure will be that appropriate to the locally most unstable spatial mode. But away from the central region, $\alpha(\omega^*)$ and ω^* will have values with large imaginary components. The appropriate eigenfunctions are then somewhat different from those associated with the most unstable mode at the centre. The slight differences in the flow structure arise mainly through the relative phases of $\phi(y)$ and $\phi'(y)$, and this leads to significant changes in the Reynolds stresses. This is illustrated by the results of computations of the eigenfunctions for a range of values of t at a particular location of interest in the experimental work which is to be discussed. Figure 4 shows the time history of the packet as it passes through location x=80 cms, (x_0 is at 20cms), and the eigenfunctions for the three positions (a), (b) and (c) are plotted on figure 5. The eigenfunctions and their derivatives are similar in each of the three cases except for the small phase differences that result in a substantially different stress function defined as follows:

$$u' = \mathcal{R}\{\phi'(y) \exp i Qt\}$$
$$\text{and } v' = \mathcal{R}\{i\alpha\phi(y) \exp i Qt\} \qquad (12)$$

In terms of A, the amplitude of the fundamental

$$u'v' = S(y,t)\, A^2(t)_x (1 + \cos 2Q_r t) \qquad (13)$$

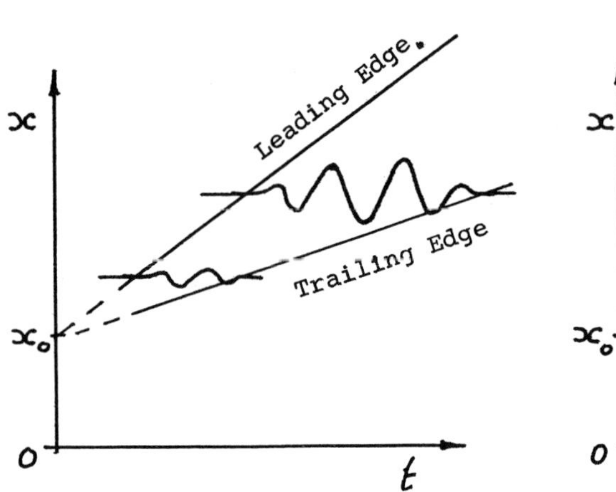

FIG. 1. Wave Packet Evolving in a Boundary Layer

FIG. 2. Disturbance Growing Exponentially

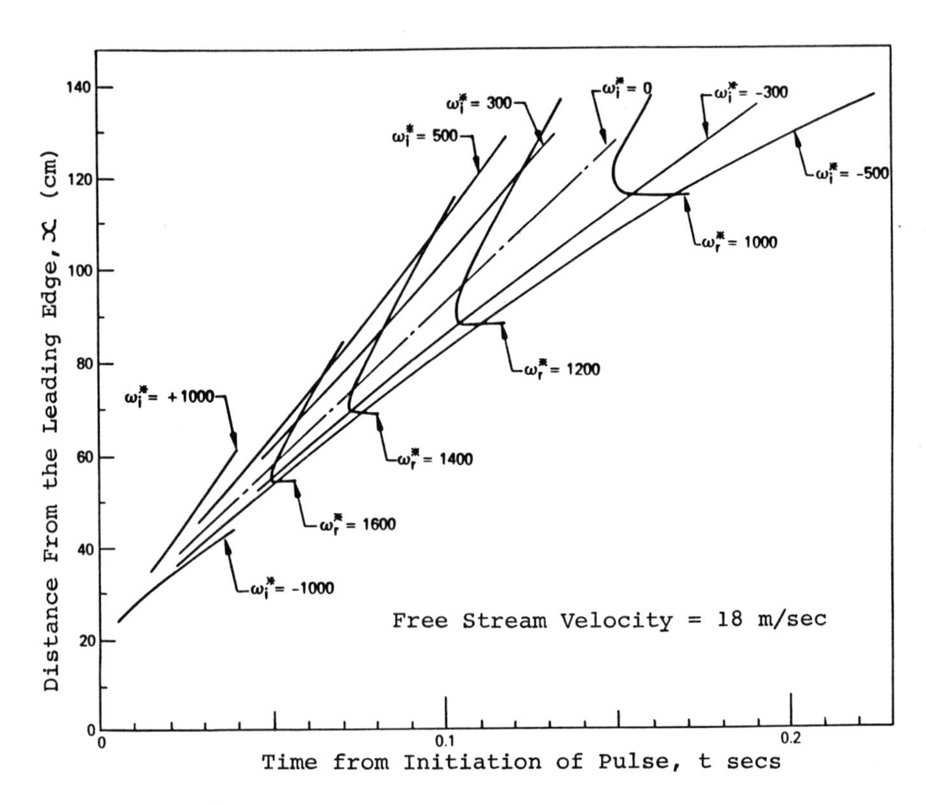

FIG. 3. Values of ω* that satisfy equation (11)

where

$$S(y,t) = \frac{\phi_r'}{2}\left[\alpha_r \phi_i + \alpha_i \phi_r\right] - \frac{\phi_i'}{2}\left[\alpha_r \phi_r - \alpha_i \phi_i\right] \tag{14}$$

There is a "mean" component as well as a second harmonic of the fundamental wave, which vary slowly with time through the packet. The non-linear terms that were neglected in the process of linearization involve the above quantity and its derivatives. We can thus make some assessment of the magnitude of these additional terms from calculations based soley on linear theory. In particular the integral through the boundary layer

$$\hat{S}(t) = \int S(y,t)\,dy \tag{15}$$

provides a measure of the neglected terms as a function of t. Now a purely periodic wave of the most unstable spatial mode will have the same eigenvalues as the mode at the centre of the wave packet. This quantity multiplied by $A^2(t)$, the square of the amplitude, is plotted as a dotted line on figure 6 for comparison with the values of S(t) computed from the asymptotic solutions. The violent "N-wave" character of the stress induced by a wave packet certainly suggests that the influence of the non-linear terms will be more important than for periodic waves, and this possibly provides the explanation for some of the observations that will be discussed next.

5. EXPERIMENTAL OBSERVATIONS OF WAVE PACKETS

I now want to briefly describe some experimental work on wave packets. The programme of work concerns the evolution of wave packets in a laminar boundary layer on a flat plate. Some of this research has already been reported [14,15], but this is intended to give a short summary of the finding for those unfamilar with the subject. The experiment was concerned with the evolution of three-dimensional wave packets excited by an acoustic driver through a small hole in the plate. Each time a pulse was introduced a wave packet was generated, and a record of its passage past a fixed probe

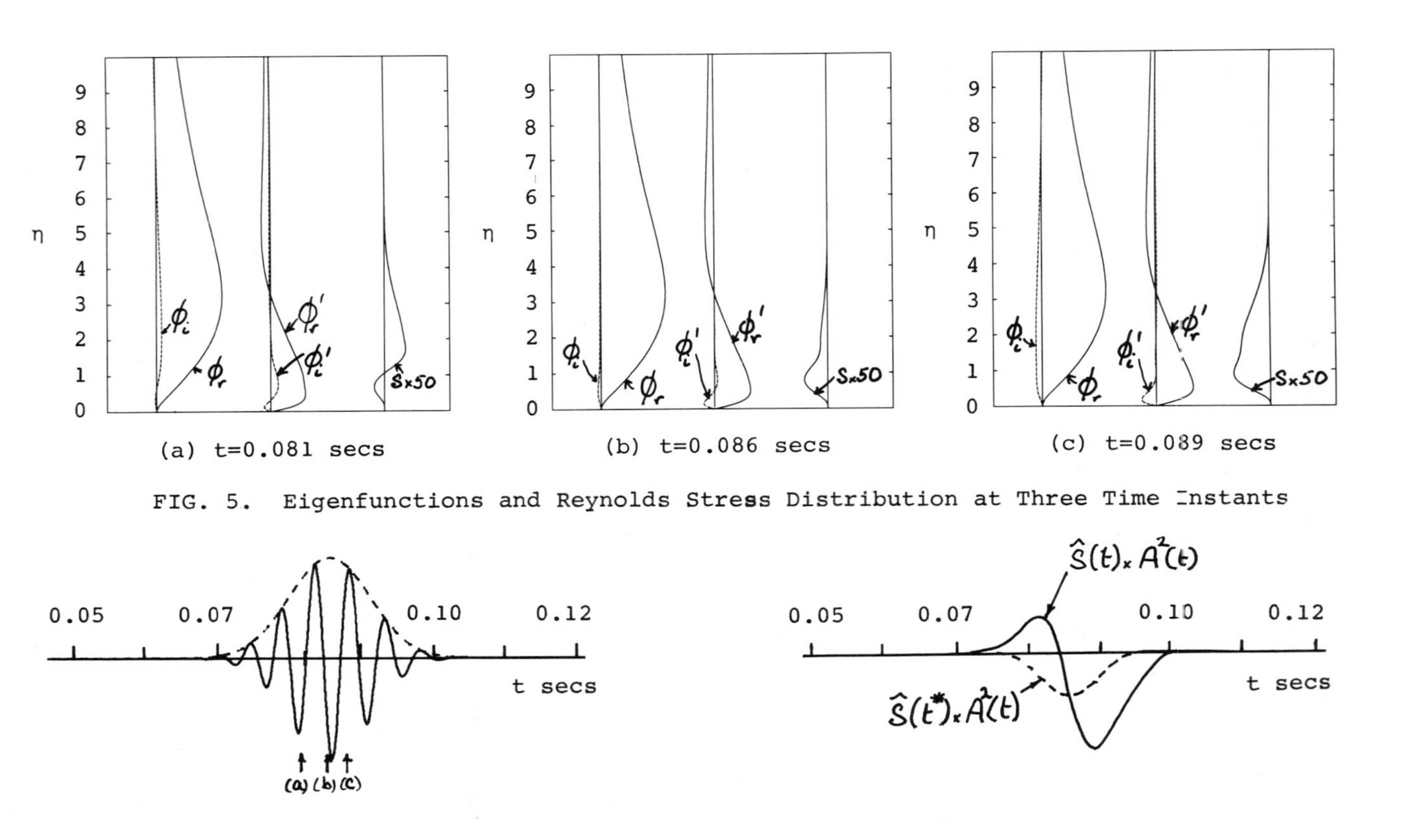

(a) t=0.081 secs

(b) t=0.086 secs

(c) t=0.089 secs

FIG. 5. Eigenfunctions and Reynolds Stress Distribution at Three Time Instants

FIG. 4. Time History of the Velocity Fluctuations in a Wave Packet

FIG. 6. Reynolds Stress in a Packet

located downstream was obtained. Random noise in the signal was virtually eliminated by forming the ensemble average of a number of separate realizations. From the data acquired at a number of equi-spaced locations across the span of the plate it has been possible to construct a perspective display (figure 7) of the streamwise velocity perturbation as a function of spanwise co-ordinate and time. Linear theory provides the behaviour of all the discrete modes that are excited, and this information can be used to compute the complete summation and so generate a model of the packet. This information is displayed on figure 8 on the same basis as that from the experiment. The experimental results follow very much the same pattern of development as predicted by this model. Initially, at least, the actual wave crest shapes were also predicted with remarkable accuracy, but far downstream the observed wave packet developed distorted wave fronts not at all predicted by these computations.

Later experiments, that have not yet been fully reported [16], confirmed these observations, and showed that the distortions in the wave fronts were associated with non-linearities. This came as something of a surprise, at the time, since the mean peak-to-peak amplitude recorded in the packet experiment was well below levels for which any non-linear effects were expected for purely periodic wavetrains. It did not seem reasonable to expect the wave packet to exhibit non-linear phenomena at lower amplitudes than the periodic wavetrain. However, it is now clear that the modulation of the basic wavetrain plays an important role in the non-linear evolution of the wave system.

6. DISCUSSION

So far, then, we have discussed the manner in which a laminar boundary layer can support waves that increase in amplitude as they evolve downstream. These waves are generated by external factors, and they therefore arise as forced oscillations rather than through naturally evolving instabilities. The process of modal selection is much more powerful in the true instability situation where even a very

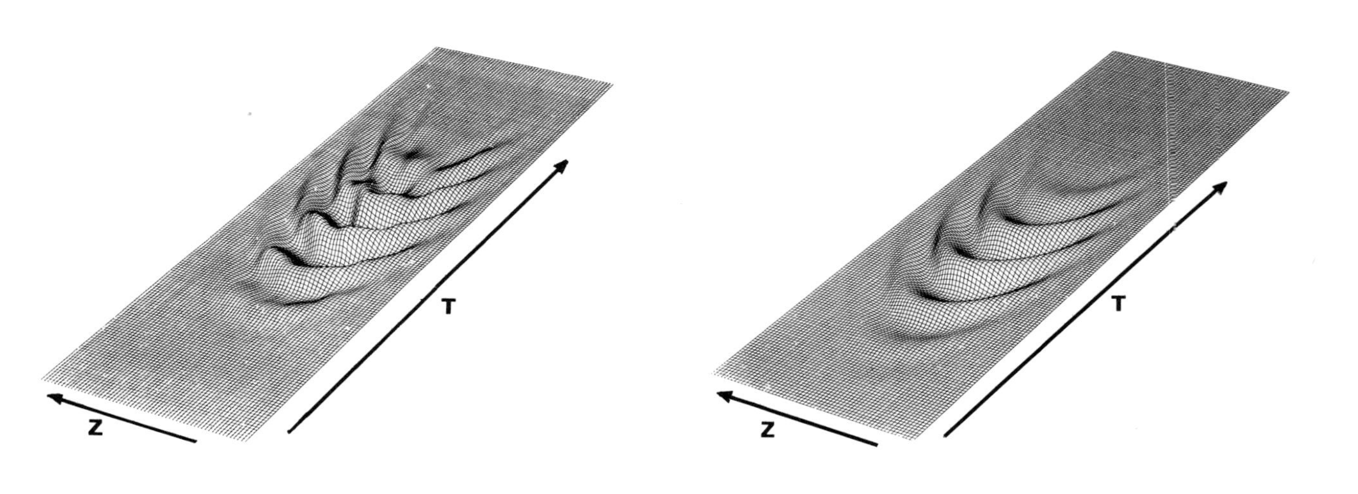

FIG. 7. Experimental Wave Packet

FIG. 8. Theoretical Wave Packet

weak initial random disturbance will become filtered and amplified so as to produce an isolated mode just above the critical Taylor number. In the boundary layer type of problem the random excitation will become filtered to some degree by the process of selective amplification, but the motion at any downstream location will, at most, be a narrow-band of waves. It can be argued that even in convective flows that support spatial waves, if the flow were truly parallel, then just above the critical Reynolds number a discrete disturbance would appear far enough downstream. However, we are concerned with the growing boundary layer where the modal selection process varies with the streamwise location, and this forces us to consider the excitation and the region over which selective amplification has taken place. For the specific case of a wavemaker driven with simple harmonic motion the appropriate form of solution far downstream is given by a spatial mode defined by an eigenvalue with a real frequency parameter coupled with a complex wavenumber, where the imaginary component accounts for the spatial growth.

When the flow is excited by an isolated pulse the streamfunction far downstream can best be modelled by a mode which has complex values of both wavenumber and frequency defined in a particular way at each position and instant of time. It turns out, therefore, that there is no simple answer to the question as to whether the waves observed in boundary layers are of the spatial or temporal form. In general it seems likely that the structure of the stream function at any particular spatial position and any given time can best be represented by a mixed mode that will depend on the form of excitation. It is not just the eigenvalues, and hence the signatures measured by a hot-wire anemometer, that depends on the type of mode, but more significantly we find that the structure of the solution in the direction normal to the boundary is also affected. In the wave packet example it transpires that the Reynolds stresses acting within the flow are greatly affected by the form of mode, and I believe this offers part of the explanation for the observed non-linear behaviour. If we want to understand the non-linear processes that arise in naturally excited flows, then the powerful effects produced by the modulation of the wavetrain need to be studied. Investigations of purely periodic waves, although interesting, do not apply directly to the real problem.

Although experiments on periodic waves are not going to unravel all the mysteries of the transition to turbulence, there are certainly some interesting phenomena worth studying. I want to briefly mention some work that I am doing on such waves. The aspect of interest concerns the way a periodic motion can evolve into broad-band random motion. It is clear that in the boundary layer the are a number of possible mechanisms for this. In some recent experiments using a periodic signal to drive the speaker, power spectra of the velocity fluctuations were measured at a number of streamwise stations, covering a wide Reynolds number range (850-2000). It was found that initially a line spectrum developed consisting of the fundamental, and a small amount of second and third harmonic. Downstream where the magnitude of the signal became large the fundamental appeared to develop side-bands. At this stage the signal itself produced an irregular trajectory on the phase-plane, reminiscent of the response seen when a non-linear system is forced off resonance. A continuous spectrum then rapidly developed and replaced the discrete components. This process took place before any high frequency bursts of turbulence occurred. These observations may well be consistent with the observations that have been made of similar processes in the Taylor problem, but it must be remembered that in the present experiments the fundamental oscillation is a forced motion. It is intended to try to identify a suitable model equation with the observed characteristics. Hopefully, such a model equation will also be capable of describing the non-linear behaviour that has been observed in the wave packet.

REFERENCES

1. Di Prima, R. C., Non-linear hydrodynamic stability. Proceedings of the Eight U.S. National Congress of Applied Mathematics, University of California, Los Angeles 1978, Western Periodicals Co. (1979).

2. Fenstermacher, P. R., Swinney, H. L., and Gollub, J. P. Dynamical instabilities and the transition to chaotic Taylor vortex flow. J. Fluid Mech. 94 103-128, (1978).

3. Lord Rayleigh, On the stability or instability of certain fluid motions, Proc. Lond. Maths. Soc. 11 57, (1880).

4. Lord Rayleigh, On the stability or instability of certain fluid motions, Proc. Lond. Maths. Soc. 19 67, (1887).

5. Schlichting, H., Zur Entstehung der Turbulenz bei der Plattenstromung. Nachr. Ges. Wiss. Gottingen, Maths-Phys. Klasse, 181-208, (1933).

6. Lin, C. C., On the stability of two-dimensional parallel flows. Quart. Appl. Maths. 3 117-142, 218-234, 277-301, (1945).

7. Schubauer, G. B., and Skramstad, H. K., Laminar boundary layer oscillations and transition on a flat plate, NACA Report 909 (1949).

8. Gaster, M., The role of spatially growing waves in the theory of hydrodynamic stability. Progress in Aeronautical Sciences Vol. 6, (1965).

9. Gaster, M., A note on the relationship between temporally increasing and spatially-increasing disturbances in hydrodynamic stability. J. Fluid Mech. 14 pp 222, (1962).

10. Gaster, M., The development of three-dimensional wave packets in a boundary layer. J. Fluid Mech. 32 pp 173, (1968).

11. Boutheir, M., J. Mecanique, 11 pp 599, (1972).

12. Gaster, M., On the effects of boundary layer growth on flow stability. J. Fluid Mech. 66 pp 465, (1974).

13. Gaster, M., Series representation of the eigenvalues of the Orr-Sommerfeld equation. J. Computational Phys. 29 No.2, (1978).

14. Gaster, M. and Grant, I., An experimental investigation of the formation and development of a wave packet in a laminar boundary layer. Proc. Roy. Soc. A. 347 pp-253, (1975).

15. Gaster, M., A theoretical model of a wave packet in the boundary layer on a flat plate. Proc. Roy. Soc. A. 347 pp 271, (1975).

16. Gaster, M., The physical processes causing breakdown to turbulence. 12th Naval Hydrodynamics Symposium, Washington, D.C. (1978).

This programme is supported by the U.S. Air Force (AFOSR Grant-80-0272); The Ministry of Defence, Royal Aircraft Establishment, Farnborough; The Admiralty Underwater Weapons Establishment, Portland, and the Department of Industry. Some of the computations were carried out at Flow Industries Inc. under AFOSR Contract F49620-78-C-0062.

National Maritime Institute,
Teddington, Middlesex,
England.

Wall Phenomena in the Final Stage of Transition to Turbulence

M. Nishioka, M. Asai, and S. Iida

1. INTRODUCTION.

In a laminar-turbulent transition, which is initially controlled by Tollmien-Schlichting (T-S) instability, a sequence of events takes place as first identified in boundary layers by Klebanoff and his co-workers in 1962 [1]. As the primary T-S wave grows, it undergoes spanwise distortions into high-frequency vortices at the peak positions. 18 years later, our understanding of the processes involved in the Klebanoff breakdown is far from complete.

We have been working on the linear and non-linear instabilities of a plane Poiseuille flow [2,3]. At the 1979 Stuttgart Symposium, we reported details of the initial stage of the wave breakdown [3]. Here, we would like to present our recent experimental results on the later stages of the breakdown and a comparison with those of coherent motions detected in a turbulent pipe flow.

2. EXPERIMENTAL PROCEDURE

A schematic diagram of the experiment is given in Fig. 1. At the top is shown the test section of a 6 m long rectangular channel, whose half-depth h is 7.3 mm and aspect ratio is 27.4. The parabolic flow comes from the left. The background turbulence is less than 0.05% of the center-line

ISBN 0-12-493240-1

velocity U_c. The primary wave is introduced by the vibrating ribbon shown near the lower wall. The orifice-loudspeaker arrangement permits artificial exitation of the secondary instability; it was used in our previous work [3] but is not utilized in the results presented here. Hot-wire probes measure the streamwise mean and fluctuating velocities, U and u. The Y-axis is measured from the lower wall and is scaled with the half-depth h. The probes move continuously in the Y- and spanwise Z-directions. The wave-growth in X is simulated at a fixed X observation position by varying the amplitude of the ribbon current. As the current increases, successive stages of the wave development are realized. Periodic sampling provides the ensemble averaging for the wave form of the u-fluctuation. The synchronizing pulse is taken from the ribbon current. The number of samples is 1024.

3. RESULTS and DISCUSSION

The whole experiment is made at a center-line velocity U_c of 9.7 m/s. This corresponds to a Reynolds number 5000, which is about 86% of the theoretical critical Reynolds number for linear instability. The frequency of the primary wave is fixed at 72 Hz. When the primary wave exceeds a

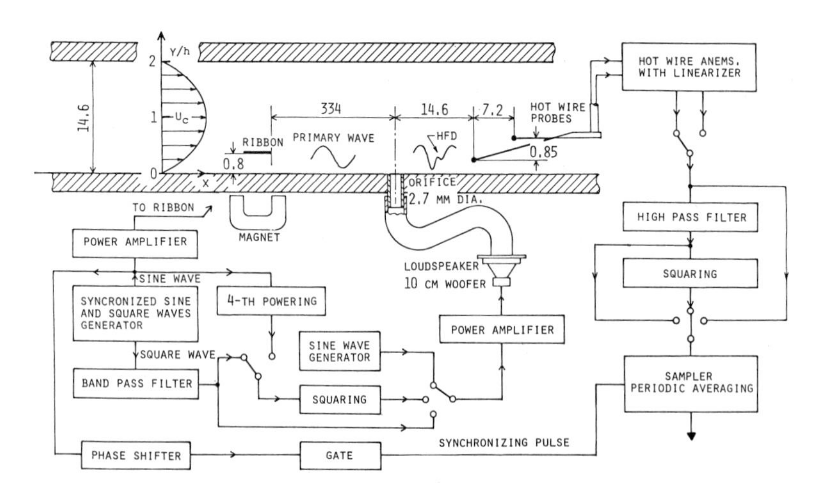

Fig. 1. Schematic diagram of experiment. Scale in mm.

threshold amplitude of about 1% of U_c, subcritical instability takes place. Subsequently, the initial very slight spanwise distortion of the wave front starts to be intensified and develops into the peak-valley structure. This is associated with a corresponding spanwise variation of U_c.

The flow system with the peak-valley structure causes local shear layers to form away from the wall at the spanwise peak positions. Essentially the same flow development was first observed for Blasius flow by Klebanoff [1] and confirmed by Kovasznay [4] and Hama [5]. Figure 2 displays u-fluctuations at a typical peak position for conditions most illustrative of the wave development. When the observed

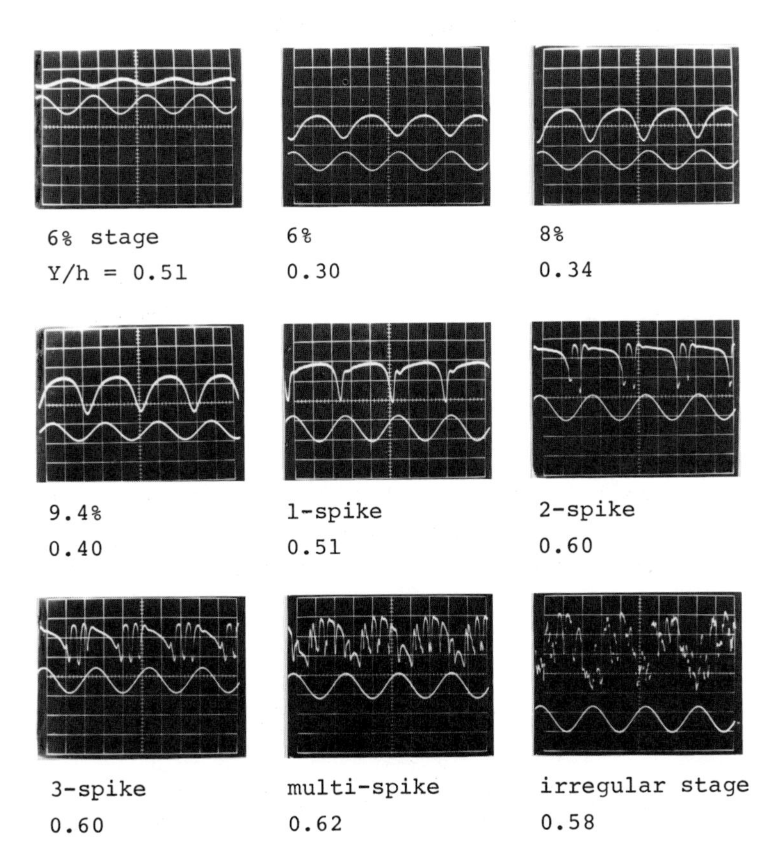

Fig. 2. Development of u-fluctuation. Traces in each photo., from top: u-fluctuation, ribbon current (72 H_z).

Y maximum in the u-fluctuation is 6% of U_c, the wave form appears sinusoidal at Y/h = 0.51, but takes on a V form near Y/h = 0.30. This V-shaped wave form is a manifestation of the local shear layer, which corresponds to the sharper trough in the wave form. The shear layer propagates faster than the surroundings [3]. This causes the downward slope in the V to increase as the wave develops downstream; this we illustrate for the 9.4% stage. Secondary instability then manifests itself by the appearance of the so-called spike; this is observed at 10.5 to 11% stages. As the amplitude increases further, the number of spikes in each cycle increases. We see in rapid succession, the 2-spike stage, the 3-spike stage, the 4, 5 and multi-spike stages. Eventually an irregular higher-frequency fluctuation appears, which can be identified with a tertiary instability. The occurrence of tertiary instability has been reported by Gaster [6] and by the present authors [3]. In the present experiment, the process is investigated through the 5-spike stage.

Figure 3 illustrates the formation of the local high-shear layer for the 9.4% stage, by means of equi-shear lines, indicating instantaneous total $\partial(U+u)/\partial Y$ non-dimensionalized with U_c/h. Time t, the abscissa runs to the left so as to provide a near-simulation of a right traveling flow pattern in X. The ordinate is the distance from the wall scaled with the half-depth h. For over one-third of the primary period T, a high-shear layer of about 0.1h in thickness, forms near Y/h = 0.4. This high-shear layer is quite

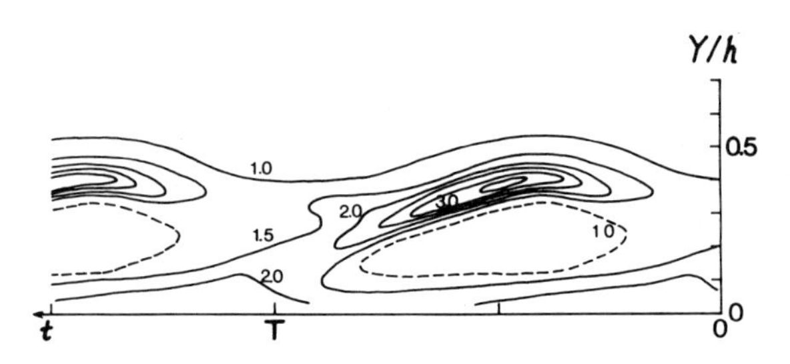

Fig. 3. $(h/U_c)\partial(U+u)/\partial Y$ = const. lines at 9.4% stage.

unstable with respect to high-frequency disturbances. That is, the secondary instability can set in.

Due to the secondary instability, the high-shear layer breaks down into discrete eddies of a scale one order of magnitude smaller than that of the primary wave. Figure 4 describes the initial stage of the breakdown, namely, the one-spike stage. The kink at the right of the high-shear layer corresponds to the beginning of the vortex roll-up process. A hot-wire senses a large low-velocity pulse when the kink passes across it.

Figure 5 shows the $\partial(U+u)/\partial Y$ pattern at 3-spike stage. The kinked portions shown in Fig. 4 have already grown into Klebanoff-type hairpin eddies, which are lifted toward the channel center and are carried downstream with increased speed. As we reported in Stuttgart[3], the hairpin eddies are subjected to tertiary instability leading to another jump to higher frequencies. Because of this, it is sometimes said that the successive high-frequency instabilities cause the flow at the peak positions to erupt into the

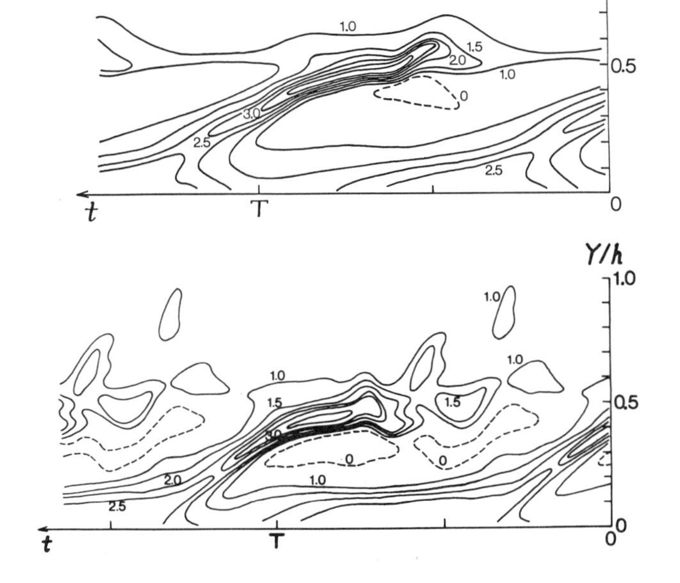

Fig. 4. Equi-shear lines at one-spike stage.

Fig. 5. Equi-shear lines at 3-spike stage.

turbulent spots of Emmons [7]. However, these successive high-frequency instabilities occur away from the wall. On the other hand, as recent investigations of wall turbulence show, turbulence is generated mainly close to the wall. Therefore, it may not be unreasonable to expect that in the final stages of transition there could occur additional wall phenomena, which could bear resemblance to those in fully developed turbulent flow. In this connection, attention should be paid to the newly formed thin shear layer near the wall at the 3-spike stage, which appears almost as a geometrical extention of the high-shear layer of the preceding cycle.

Figure 6 displays the ensemble-averaged $\partial(U+u)/\partial Y$ pattern at the 5-spike stage, again at the peak position. The uppermost line has no physical meaning other than to indicate the periodicity and the skewing feature of the development. The fast traveling first-born eddy is now located at the channel center above the high-shear layer of the preceding cycle. The topography of these instantaneous $\partial(U+u)/\partial Y$ loci is getting rather complex. The whole pattern is not inconsistent with that of the large-scale coherent motions in a turbulent boundary layer. However, any deeper similarity must be documented through other characteristics. The most significant feature probably is the development of the thin shear layer close to the wall mentioned in connection with Fig. 5. It intensifies and seems to

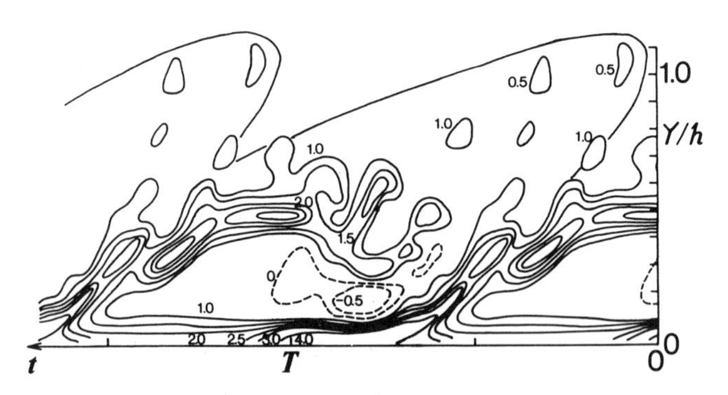

Fig. 6. $(h/U_c)\,\partial(U+u)/\partial Y$ = const. lines at 5-spike stage.

form local concentrations of vorticity on a scale of the secondary instability. These new formations must be reflected in statistical quantities such as the time mean velocity and the root mean square intensity of fluctuation.

In Fig. 7, the mean velocity at the 3-spike stage is shown on a logarithmic plot of U^+, i.e. U/u_τ against y^+, i.e. Yu_τ/ν. The broken line is the usual law of the wall. Clearly, the distribution deviates from the log-law at this stage. Figure 8 shows the time mean velocity at the 5-spike stage. The change from Fig. 7 suggests gradual approach to the log-distribution near the wall. In concert with this fact, the Y-distribution of r.m.s. distribution begin to develop a second peak close to the wall, as shown in Fig. 9. The original maximum remaining near Y/h = 0.4 is due to the

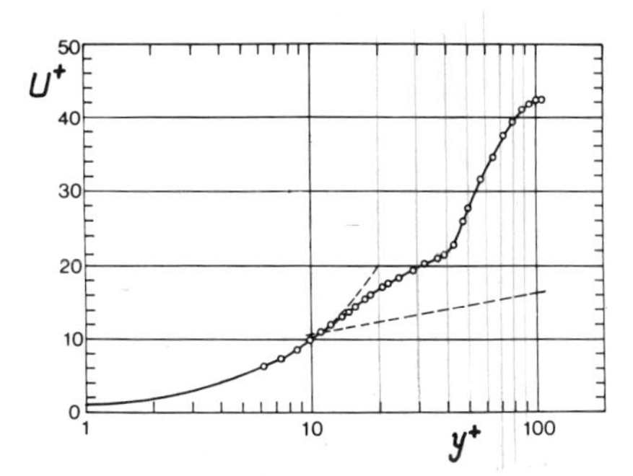

Fig. 7. Mean velocity at 3-spike stage.

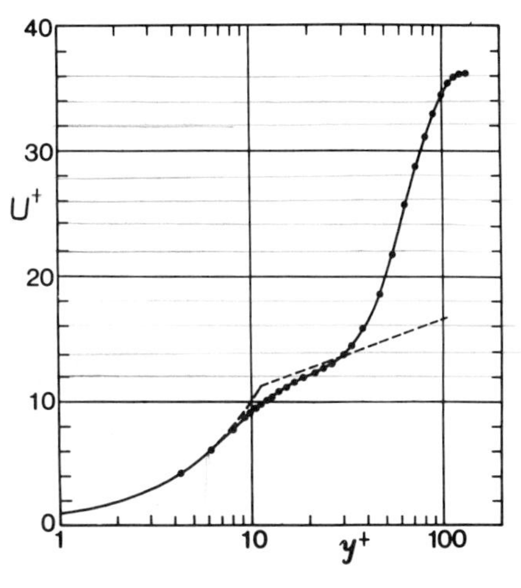

Fig. 8. Mean velocity at 5-spike stage.

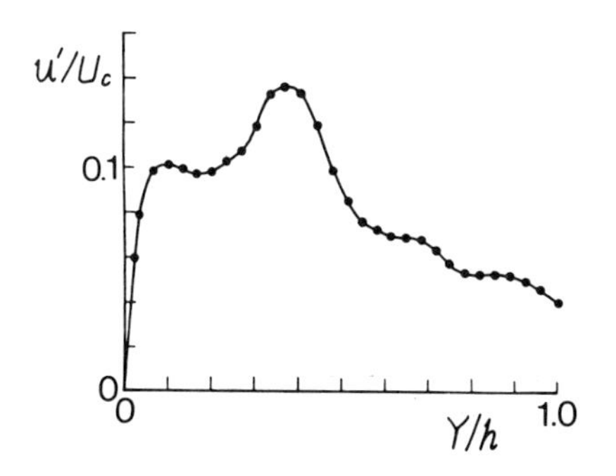

Fig. 9. Y-distribution of r.m.s. u-fluctuation at 5-spike stage.

high-shear layer. This outer maximum disappears as the randomness of the complex vorticity interactions increases.

Thus far, the development of the secondary instability has been examined on the basis of the equi-shear lines. Most interesting in this almost periodic flow is the evolution of the mean velocity distribution toward the log-law. Of course, the entire flow is not exactly periodic. However, the flow is sufficiently regular so that our periodic sampling provides accurate wave forms of the primary and secondary motions. Figure 10 compares an ensemble average with 3 samples of instantaneous wave forms. The total number of samples is 1024. The comparison shows that the periodic sampling indeed recovers the wave form almost completely as far as the primary and secondary disturbances are concerned. Here, it should be added that when the hairpin eddies near the channel center (see Fig. 6) pass across a hot-wire, it senses very sharp spikes just like as those observed by Klebanoff et al [1].

Figure 11 shows ensemble-averaged u-fluctuations at various Y/h positions at the peak. The $\partial(U+u)/\partial Y$ equi-

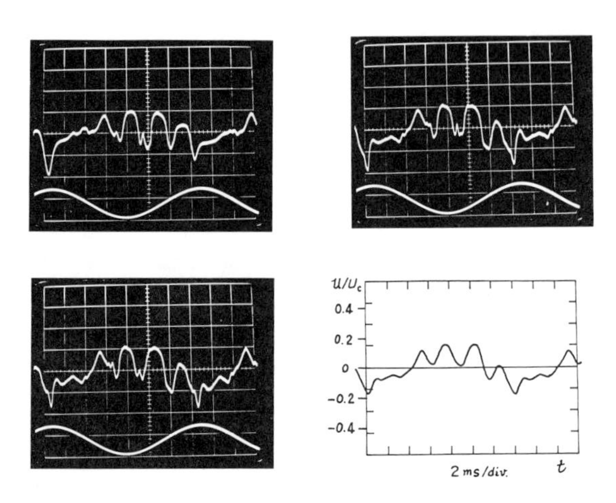

Fig. 10. A comparison between instantaneous and ensemble-averaged wave forms at the 5-spike stage, at Y/h = 0.50.

shear lines previously shown in Fig. 6 are obtained from these wave forms and the mean velocity distribution. One may recognize the thin shear layer close to the wall from these data. A strong acceleration occurs from a state of negative fluctuation to that of positive fluctuation, and this takes place earlier in time farther away from the wall. Then we see the shear layer formation at a time instant t/T = 0.5 to 0.6, across Y/h = 0.1 to 0.3. This feature has been clearly demonstrated in Fig. 6, by means of equi-shear lines.

Concerning the shear layer, it is interesting to see the spanwise behaviour at t/T = 0.5. In Fig. 12, the solid lines represent the instantaneous u-velocity distribution in the spanwise direction at various heights. Broken-lines represent corresponding mean velocity distributions. The peak position on which we have focused thus far is located at Z = 74.5 mm. At the beginning of the non-linear wave development, the spanwise distance between peak positions is 20 to 24 mm. At this 5-spike stage, the characteristic spanwise scale associated with the new shear layer is much

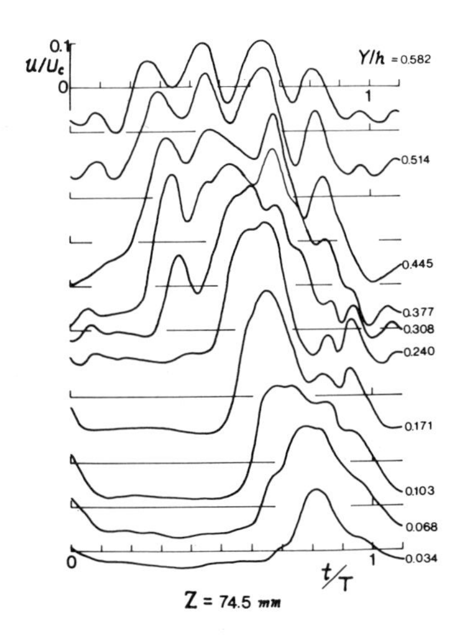

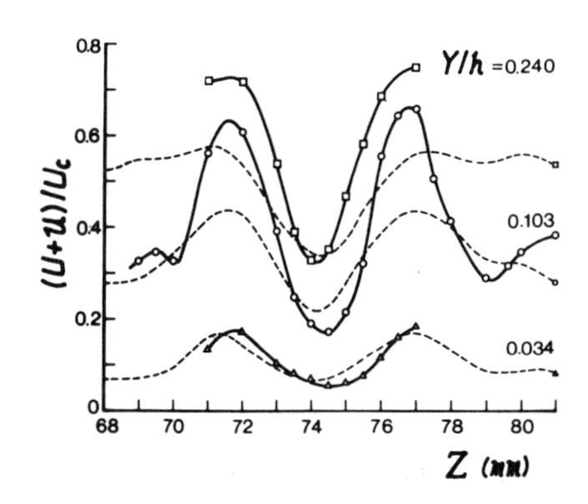

Fig. 12. Spanwise distribution of (U+u) at t/T = 0.5, at 5-spike stage.

Fig. 11. Ensemble-averaged wave forms at the 5-spike stage.

smaller, approximately 5 mm. This corresponds to $\Delta Z^+ = 80$, when scaled with u_τ/ν. The spanwise characteristic length λ^+ in wall turbulence is about 100. It appears that at the 5-spike stage we observe motions with spanwise scales which approach those of the smallest coherent structure in fully developed wall turbulence. In Fig. 13, shear layers close to the wall at Z = 74.5, 75.5 and 76.5 mm, at the 5-spike stage are compared. These results suggest that the wall shear layer may be associated with a horse-shoe vortex not unlike Theodorsen's [8], shown schematically at the bottom.

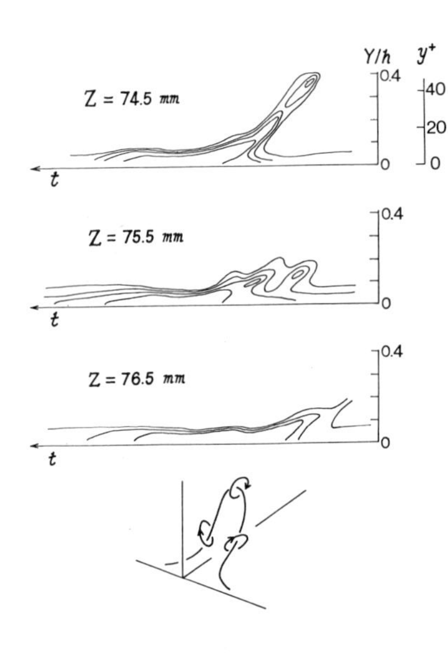

Fig. 13. Shear layers close to the wall at 5-spike stage.

Now, we look at some aspects of wall turbulence. One of the important characteristic features in wall turbulence is the repeated occurrence of strong acceleration from large negative to positive values of u(t) as clearly demonstrated by Blackwelder and Kaplan [9]. We have examined this feature in a turbulent pipe flow. First the detection scheme should be explained. Any instantaneous trace of u-fluctuation will have acceleration portions such as schematically shown in Fig. 14, a rise from $-\alpha u'$ to $\alpha u'$ in a time segment t_c. From traces detected at y^+ of 15, we selected and counted the numbers of occurrence of such accelerations for two kinds of α, and for various values of t_c in a preliminary examination. For $\alpha = 1.5$, u(t) rises continuously from negative to positive u by at least 3 times the r.m.s. value u', and for $\alpha = 1.8$ by at least 3.6u'. We also selected samples of the reverse, namely the deceleration of the same magnitude during the same observation time. N_+ and N_- are the numbers of occurrence of such accelerations and decelerations, respectively. In Fig. 14, the variation of N_-/N_+ with the nondimensional time t_c^+, i.e. $t_c u_\tau^2/\nu$ is shown for Reynolds numbers 6000 and 60000. For small values of t_c^+ of 10 or less, N_- does not exceed 10% of N_+, independently of our threshold. This means that there are virtually no strong decelerations but primarily accelerations.

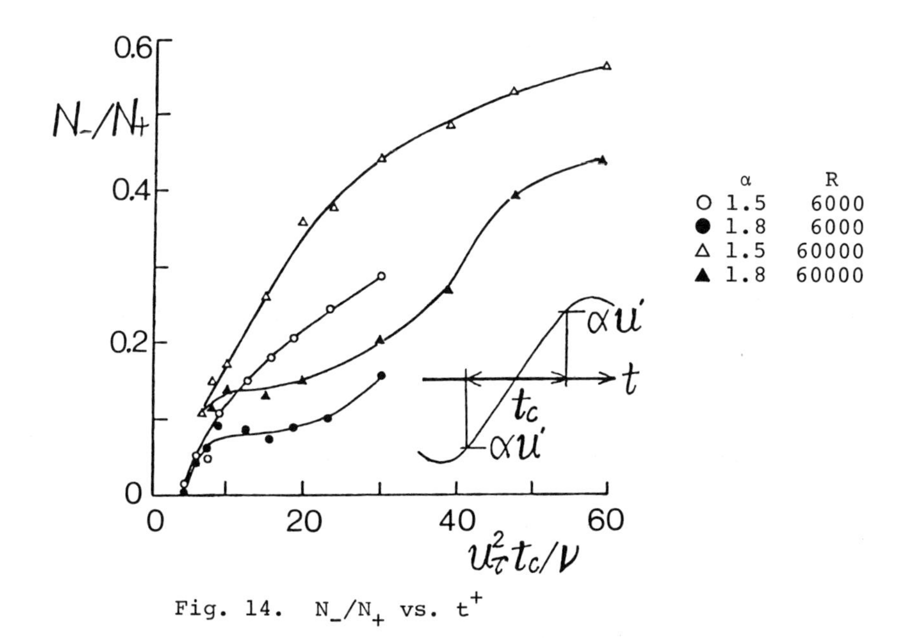

Fig. 14. N_-/N_+ vs. t^+

Figure 15 shows ensemble-averaged wave forms at various y^+ positions in the pipe. The detection is made by setting $\alpha = 1.5$. The non-dimensional time t_c^+ and the number of samples are respectively 6 and 32 for R = 6000, and 9.8 and 128 for R = 60000. With the detector probe at $y^+ = 15$, the

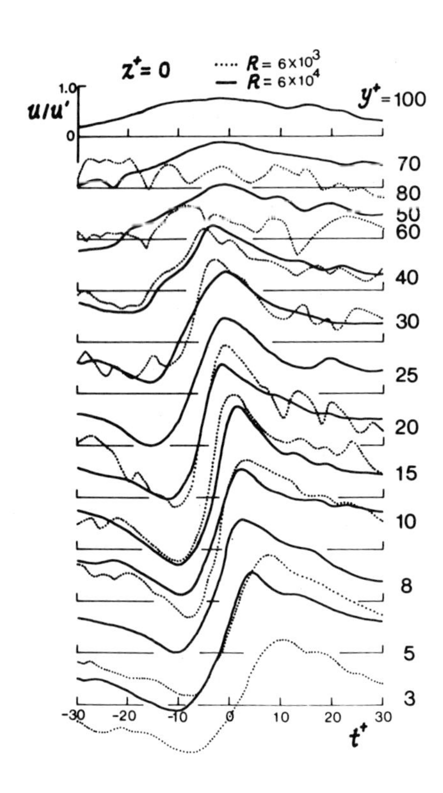

Fig. 15. Ensemble-averaged u-fluctuations of coherent motions in a turbulent pipe flow.

sampling probe gave us the ensemble-averaged wave forms for various heights y^+. The wave forms for R = 6000 are richer in high-frequency content because the number of samples is only a quarter of that for R = 60000. Even this difference tends to disappear close to the wall. Thus, in this non-dimensional representation, there is no essential difference due to Reynolds number. Moreover, these wave forms are quite similar to those shown in Fig. 11 for the 5-spike stage in the wall region. Thus, patterns of high acceleration at this late stage of the breakdown appear to approach the characteristics of the wall turbulence; we also found that λ^+ of the coherent motions shown in Fig. 15 is about 80 for both Reynolds numbers.

Figure 16 shows equi-shear lines obtained from the

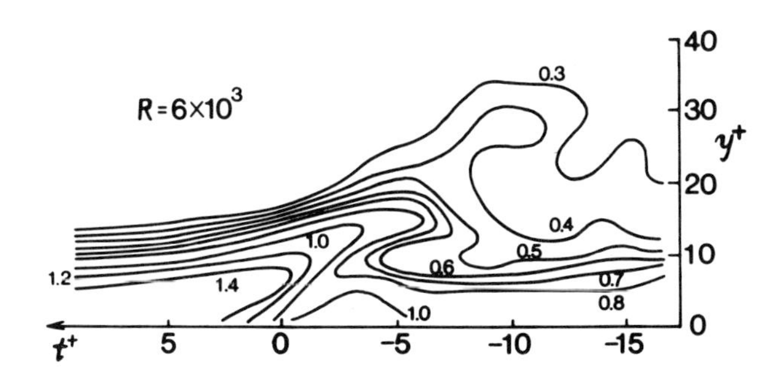

Fig. 16. Equi-shear lines in a turbulent pipe flow; $\partial(U^+ + u^+)/\partial y^+$ = const. lines.

previous turbulent pipe data for R = 6000. The abscissa, time t^+ is running in the left direction. Needless to say, the resemblance between the patterns in this fully developed turbulence and those at the 5-spike stage in the transition case is hardly accidental. We should add that a similar result is obtained for R = 60000, so that Reynolds number dependence of this feature is not strong.

4. CONCLUDING REMARKS

Many features of the observed wall phenomena in the later stages of the breakdown appear similar to those in the turbulent case. Could this then be called the beginning of a turbulent spot? We do not know; partly because sufficient conditions defining a nascent spot are not known. Additional features may still have to develop, but probably not far downstream.

The authors wish to express their sincere gratitude to Professor Mark V. Morkovin for his valuable advice and discussion, and also to Professors I. Tani and H. Sato for their continual encouragement.

REFERENCES

1. Klebanoff, P. S., K. D. Tidstrom and L. M. Sargent, The three-dimensional nature of boundary layer instability, J. Fluid Mech. 12 (1962), 1-34.

2. Nishioka, M., S. Iida and Y. Ichikawa, An experimental investigation of the stability of plane Poiseuille flow, J. Fluid Mech. 72 (1975), 731-751.
3. Nishioka, M., M. Asai and S. Iida, An experimental investigation of the secondary instability, in Laminar-Turbulent Transition (Proc. IUTAM Symposium, Stuttgart, 1979; R. Eppler and H. Fasel, eds), Springer-Verlag, 1980, 37-46. Also reported in Proc. 11th Turbulence Symposium, Inst. Space Aeron. Sci., Tokyo Univ. (1979).
4. Kovasznay, L. S. G., H. Komoda and B. R. Vasudeva, Detailed flow field in transition, Proc. 1962 Heat Transfer & Fluid Mech. Inst., Stanford Univ. Press, 1962, 1-26.
5. Hama, F. R. and J. Nutant, Detailed flow-field observations in the transition Process in a thick boundary layer, Proc. 1963 Heat Transfer & Fluid Mech. Inst., Stanford Univ. Press, 1963, 77-93.
6. Gaster, M., The physical processes causing breakdown to turbulence, 12th Naval Hydrodynamics Symposium, Washington, 1978.
7. Emmons, H. W., The laminar-turbulent transition in a boundary layer, Part 1, J. Aeron. Sci. 18 (1951), 490-498.
8. Theodorsen, T., The structure of turbulence, in 50 Jahre Grenzschichtforschung (H. Gorter and W. Tollmien eds), Friedr. Vieweg & Sohn, 1955, 55-62.
9. Blakwelder, R. F. and R. E. Kaplan, On the wall structure of the turbulent boundary layer, J. Fluid Mech. 76 (1976), 89-112.

College of Engineering
University of Osaka Prefecture
Sakai, Osaka
Japan

Subcritical Transition to Turbulence in Planar Shear Flows

S. A. Orszag and A. T. Patera

1 Introduction

There is a large discrepancy between the results of parallel flow linear stability theory and experimental observations of transition in plane channel flows. Experiments by Davies & White (1928), Kao & Park (1970), and Patel & Head (1969) indicate that plane Poisciulle flow can undergo transition at Reynolds numbers, R, of roughly 1000 when the background disturbance level is about 10 *per cent*. The strongly nonlinear nature of the problem is reflected in the fact that Nishioka *et al* (1975) have maintained laminar flow to $R = 8000$ by keeping the background disturbance level sufficiently low. While there have been fewer experimental investigations of plane Couette flow, available data (Reichardt 1959) suggest a transitional Reynolds number as low as or lower than that for plane Poiseuille flow.

The disparity between the experimental evidence and the results of parallel flow linear theory (which predicts a critical Reynolds number of 5772 for plane Poiseuille flow and ∞ for plane Couette flow) requires consideration of finite-amplitude effects. Nonlinear stability analysis typically involves the search for finite-amplitude equilibria, the stability of the flow being inferred by assuming the equilibria to be critical points in a one-dimensional phase space for the amplitude of the (primary) disturbance. A large number of these analyses (following the seminal work of Meksyn & Stuart 1951 and Stuart 1960) rely on amplitude expansions that have very small (Herbert 1980) or unknown radii of convergence . In plane Poiseuille flow (where a finite linear neutral curve exists) these techniques can successfully predict small-amplitude equilibria close to the neutral curve. However, upon extension to situations where a finite neutral curve does not exist (e.g. plane Couette flow, Hagen-Poiseuille flow), these techniques yield spurious equilibria.

In contrast to the amplitude expansion techniques, the iterative method due to Zahn *et al* (1974) and Herbert (1976) is able to reliably calculate large-amplitude as well as small-amplitude equilibria. Here, the nonlinear eigenvalue problem for travelling-wave solutions to the Navier-Stokes equations (rendered steady by an appropriate Galilean transformation) is solved numerically using a spectral expansion and Newton iteration. It is on the basis of results obtained using this method (Herbert 1977) and full simulations that we claim plane Poiseuille flow has both sub-critical and super-critical finite-amplitude equilibria whereas plane Couette flow (Orszag & Kells 1980; Patera & Orszag 1980a) and Hagen – Poiseuille (Patera & Orszag 1980b) have none.

ISBN 0-12-493240-1

Most nonlinear stability results obtained to date are two-dimensional. There have been investigations of low-order two- and three- dimensional interacting wave systems that exhibit resonance - induced instabilities (Benney & Lin 1960; Craik 1980; Gustavsson & Hultgren 1980), however, it is not yet clear whether these instabilities lead directly to transition. For instance, in Hagen-Poiseuille flow, near-synchronization between the primary and harmonic occurs for all (α, R), but slow secular growth is maintained for such a short time that the perturbation remains linear (Patera & Orszag 1980b).

It is generally believed that the instability leading to transition in plane channel flows is a "secondary instability" (Herbert & Morkovin 1980), i.e. an instability which acts on the combined flow consisting of the basic parallel flow and a finite-amplitude two-dimensional cellular motion. Furthermore, the three-dimensional nature of shear-flow turbulence suggests that the secondary instability should be three-dimensional. That such an instability would lead to aperiodic, apparently random behaviour upon attaining finite-amplitude rather than undergoing nonlinear saturation has been demonstrated numerically by Orszag & Kells (1980) and Patera & Orszag (1980a). Indeed, the transition process is inherently three - dimensional as shown by the fact that even initially "turbulent" two - dimensional fields relax to a laminar state. In this paper we present a three - dimensional secondary instability that predicts transitional Reynolds numbers in good agreement with experiment.

In Section 2 we describe the two-dimensional steady and time dependent properties of plane Poiseuille and plane Couette flows using the iterative technique of Zahn *et al* (1974) and Herbert (1976) and full numerical simulation of the Navier-Stokes equations. Two spectral codes (both implemented on the Cray - 1 computer) have been used to verify the behaviour reported here. The first employs a splitting technique similar to that described in Orszag & Kells (1980), while the second involves a full-step second-order in time method. Details of the latter are given in the Appendix.

In Section 3 it is shown that the finite-amplitude two-dimensional states investigated in Section 2 are strongly unstable to very small three-dimensional perturbations. Our aim here is to show (by full numerical simulation) that this explosive secondary instability can explain the sub-critical transitions that often occur in real flows.

Finally, in Section 4 , we show that the three-dimensional instability described in Section 3 can be analyzed by a linear stability analysis of a two-dimensional flow consisting of the basic parallel flow and a steady (or quasi-steady) finite-amplitude two-dimensional cellular motion.

A brief presentation of the results of the present paper was given by Orszag & Patera (1980).

2 Two-Dimensional Finite-Amplitude States

The instability to be described in Section 3 involves the rapid exponential growth of three-dimensional perturbations superposed on two-dimensional finite-amplitude flows. As the existence of the instability depends critically on the time scales associated with the evolution of the two-dimensional finite-amplitude motions, it is necessary to understand two - dimensional behavior before discussing three - dimensional effects. Just as in the linear regime where plane Poiseuille flow and plane Couette flow differ in that the former has a finite neutral curve wheras the latter does not, so in the nonlinear regime they differ in that plane Poiseuille flow admits finite-amplitude equilibria while plane Couette flow apparently does not. We begin our discussion by investigating equilibrium solutions in plane Poiseuille flow.

In plane Poiseuille flow two - dimensional finite – amplitude solutions to the Navier - Stokes equations are sought in the form of travelling waves (Zahn *et al* 1974; Herbert 1976),

$$\mathbf{v}(x,z,t)=\mathbf{F}(x-ct,z)+(1-z^2)\hat{\mathbf{x}} \tag{1}$$

where $\hat{\mathbf{x}}$ is a unit vector in the streamwise direction x , z is the cross-stream direction, and c is a real wave speed. No-slip boundary conditions hold at the walls, $z=\pm 1$, and periodicity with wavelength $\lambda = 2\pi/a$ is assumed in the streamwise direction.

The locus of points in (E,R,a) space for which a solution of the form (1) exists is called the neutral surface, where E is the energy of the disturbance relative to the basic flow $\mathbf{U}=(1-z^2)\hat{\mathbf{x}}$ and R is the Reynolds number based on centerline velocity and channel half-width, $h=1$. Such a neutral surface (as calculated by Herbert) is plotted in Fig. 1.

The stability analysis given later involves linearization about the neutral states (1), and so we summarize here our numerical methods (similar to those of Herbert 1976) for finding these solutions of the Navier-Stokes equations. If the streamwise and cross-stream velocities are written in terms of a stream-function as $u=\partial\psi/\partial z$ and $w=-\partial\psi/\partial x$,respectively, the Navier-Stokes equations become

$$\frac{\partial}{\partial t}\nabla^2\psi+\frac{\partial(\psi,\nabla^2\psi)}{\partial(z,x)}=\nu\nabla^4\psi \tag{2}$$

where $\nu = 1/R$. We look for solutions periodic in x and steady in a frame moving with speed c relative to the laboratory frame,

$$\psi=\sum_{n=-\infty}^{\infty}\tilde{\psi}_n(z)e^{ina(x-ct)}.$$

The nonlinear eigenvalue problem for $(\tilde{\psi}_n, c)$ is then given by

$$\nu(D^2-n^2a^2)^2\tilde{\psi}_n-ian([\overline{U}-c](D^2-n^2a^2)\tilde{\psi}_n-(D^2\overline{U})\tilde{\psi}_n)$$
$$+ia\sum_{m=-\infty}^{\infty}\left[(n-m)\tilde{\psi}_{n-m}(D^3-m^2a^2D)\tilde{\psi}_m-mD\tilde{\psi}_{n-m}(D^2-m^2a^2)\tilde{\psi}_m\right]=0 \tag{3}$$

where $D=\partial/\partial z$ and $\overline{U}$ is the basic (parabolic) profile. Reality of ψ requires

$$\tilde{\psi}_n=\tilde{\psi}^{\dagger}_{-n}, \tag{4}$$

where superscript † denotes complex conjugation. We also impose the symmetry requirement that $\psi(x,z)=\psi(x+\pi/a,-z)$ or

$$\psi_n(z)=(-1)^{n+1}\tilde{\psi}_n(-z) \tag{5}$$

(which is consistent with (3)), in order to reduce the computational complexity of the problem.

Two boundary conditions must be provided at each wall for each Fourier mode. Noting the symmetry condition (5) it suffices to impose

$$\tilde{\psi}_n(1) = D\tilde{\psi}_n(1) = 0 \qquad n \neq 0 \tag{6}$$
$$D\tilde{\psi}_n(1) = D^2\tilde{\psi}_n(1) = 0 \qquad n = 0, \tag{7}$$

where (6) follows from the no slip condition while (7) derives from the additional requirement of no mean disturbance stress at the wall (i.e. the mean pressure gradient remains $2/R$ as in the basic undisturbed flow).

To solve the equations (3) numerically a Galerkin approximation in terms of Fourier modes in the x-direction is used (i.e. (3) is truncated at $m = N - 1$), and a Chebyshev pseudospectral method (Gottlieb & Orszag 1977) is used in the z – direction. The resulting nonlinear algebraic eigenvalue problem is solved using a Newton iteration in conjunction with an arc-length continuation method (Lentini & Keller 1980) to avoid problems at limit points. Note that only one of the boundary conditions for each Fourier mode can be handled by replacing the dynamical equation at the wall. The second boundary condition results in an additional equation and a corresponding τ – factor (Gottlieb & Orszag 1977). The solution to (3) is unique only to within an arbitrary phase, and, upon specifying this, the numerical problem is completely determined.

Laminar equilibria of the form (1) exist down to Reynolds number $R \approx 2900$ (see Fig. 1). However, transition typically occurs (and the three - dimensional instability described in Section 3 obtains) down to $R = O(1000)$. In order to understand how an instability mechanism which requires a two - dimensional secondary flow can persist down to a Reynolds number at which all two - dimensional flows decay, one must investigate the time-scales associated with approach to equilibrium.

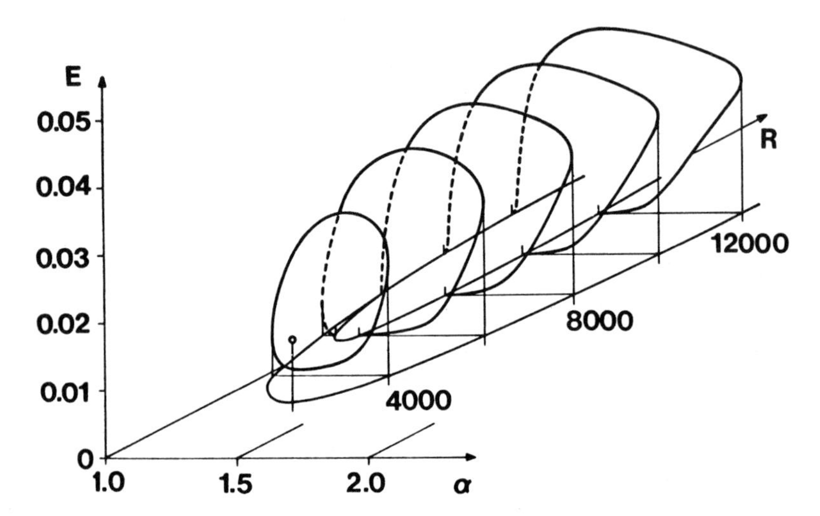

Figure 1. The neutral surface in (E, R, α) space for plane Poiseuille flow. Finite-amplitude neutral disturbances exist down to a Reynolds number $R \approx 2900$. The critical wavenumber, α_c, is shifted from the linear value of 1.02 at $R \approx 5772$ up to 1.32 at $R \approx 2900$. (Courtesy of Dr. Th. Herbert)

A sub-critical ($2900 < R < 5772$) slice of the neutral surface (Fig. (1)) is given at $R = 4000$ in Fig. 2, the upper part of the oval being termed the upper branch **(UB)**, the lower part termed the lower branch **(LB)**. An oversimplified argument based on a one-dimensional phase space representation for E correctly predicts the UB solutions to be stable and the LB solutions to be unstable (since in a one - dimensional phase space the stability of critical points typically alternates).

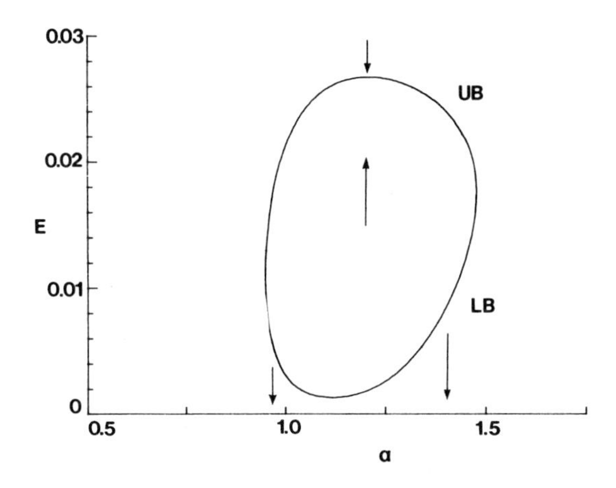

Figure 2. A subcritical (E, α) slice of the neutral surface for plane Poiseuille flow at $R = 4000$. The stability of solutions is indicated by the arrows. The behavior shown in this plot is typical for $2900 < R < 5772$.

We can determine the time scale on which flows approach UB solutions by noting that in a periodic steady flow vorticity varies by at most $O(1/R)$ along an *interior* streamline. Indeed, when $R = \infty$, the Jacobian in (2) must vanish in a reference frame moving with speed c. This result is illustrated in Figs. 3 and 4 in which streamlines of the steady flow (1) and vorticity contours, respectively, are plotted. Note the similarity of the plots in the interior of the flow. The evolution of a typical flow to the equilibrium state (1) may then be described as follows. First, on a (fast) convective time scale, the flow achieves a state in which vorticity is a function of the streamfunction to $O(1/R)$. Final approach to equilibrium then occurs on a purely diffusive time scale, of order R. (Note this suggests that if a strong secondary instability exists, it is not two - dimensional).

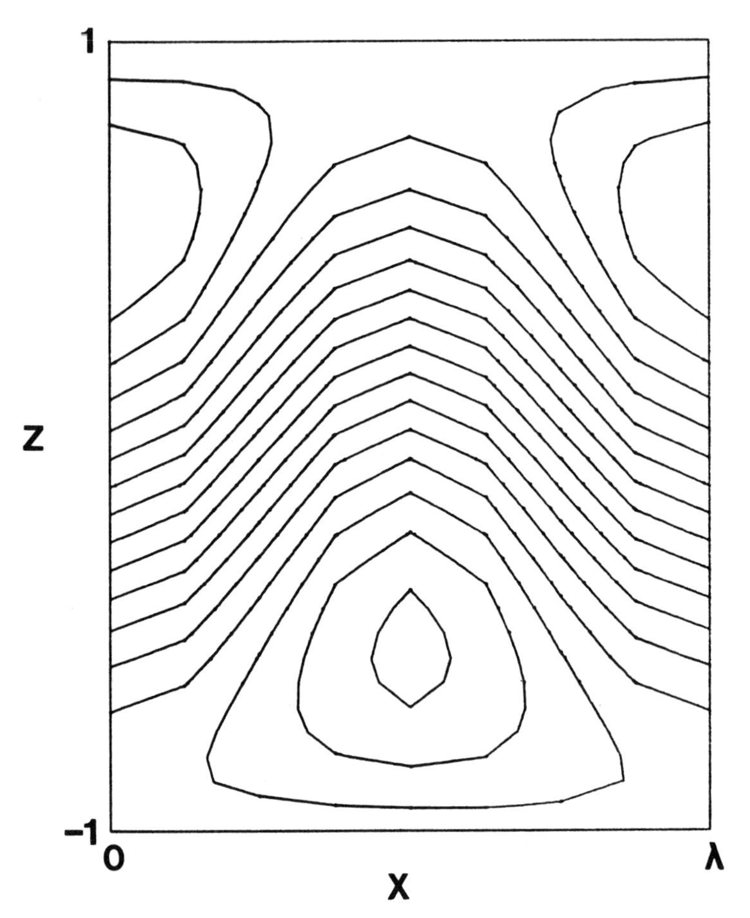

Figure 3. Streamlines of the steady (stable) finite-amplitude two-dimensional plane Poiseuille flow of the form (1) at $R = 4000, \alpha = 1.25$. The secondary motion appears as counter-rotating eddies. Here $\lambda (= 2\pi/\alpha)$ is the wavelength of the primary.

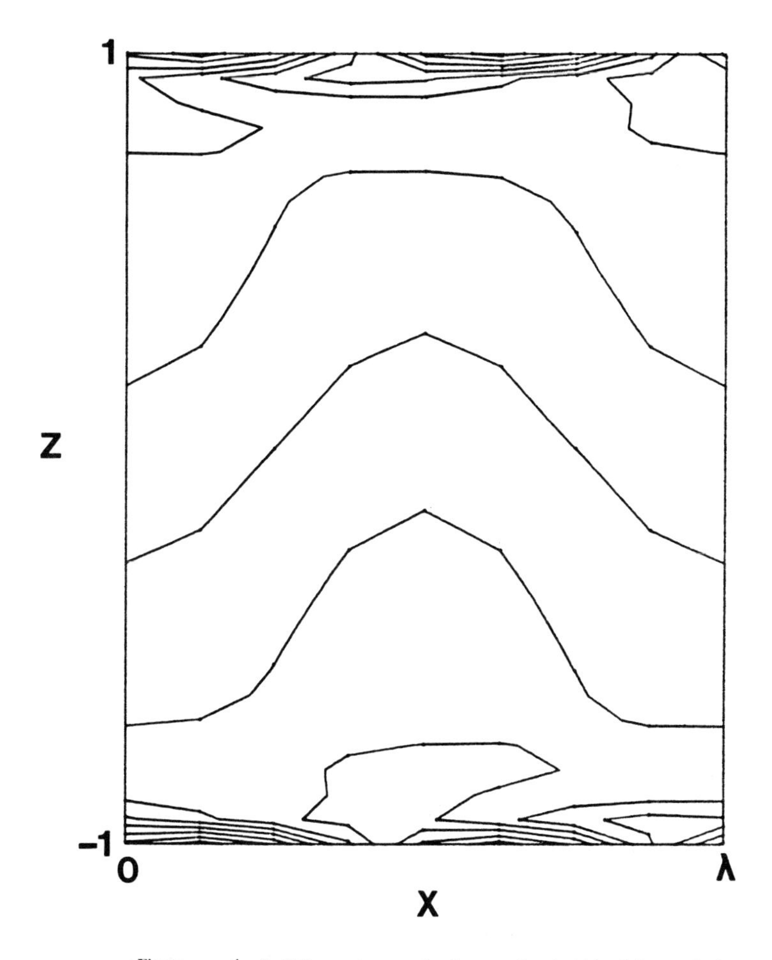

Figure 4. Vorticity contours of the steady (stable) finite-amplitude two - dimensional plane Poiseuille flow of the form (1) at $R = 4000, \alpha = 1.25$. Note that in the interior of the flow (where viscosity is unimportant), the vorticity contours are very similar to the streamlines in Fig. 3 . This implies that the nonlinear interaction is small away from the boundaries.

This prediction of diffusive approach to equilibrium is validated in Fig. 5 as a plot of the projection of numerical solutions to the two-dimensional Navier-Stokes equations on the two - dimensional phase space $(\sqrt{E_1}, \sqrt{E_2})$. Here E_n is the energy of that part of the flow which depends on x like e^{ianx}. In Fig. 5, the dots, equally spaced in time, indicate the actual evolution of trajectories emanating from different initial conditions. The arrows show schematically the orbits followed by the different flows. With regard to time scales, the important point to note is that each initial condition above the threshold energy quickly evolves on a time scale of order 10 to a state within a band of quasi-equilibria (illustrated by the shaded region in Fig. 5), and then only very slowly (indicated by the clustering of dots) approaches the steady solution on a time scale of order R. The integration times in Fig. 5 are as large as 800. A plot of E_1 and E_2 vs. t for one initial condition is given in Fig. 6, showing that appreciable deviations from the equilibrium solution only decay by cross-streamline diffusion on a time scale on the order of R. The fact that the vorticity-diffusion argument given above applies to such a wide band of phase space around the equilibrium suggests that it may also apply for nearby flows with $R < 2900$ (i.e. at Reynolds numbers where there are no finite-amplitude equilibria). This is indeed the case, as shown in Fig. 7 by the slow decay of a disturbance at $R = 1500, a = 1.32$.

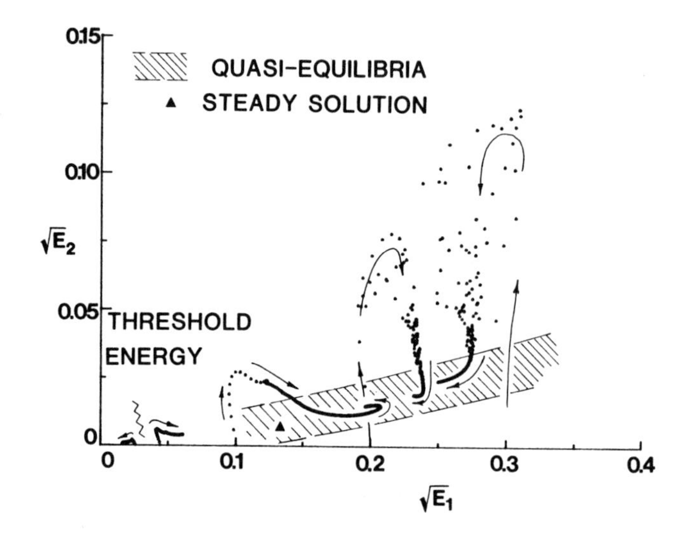

Figure 5 . A phase portrait of disturbances to laminar parallel plane Poiseuille flow in $(\sqrt{E_1}, \sqrt{E_2})$ space at $R = 4000, \alpha = 1.25$. The dots, equally spaced by 1.25 in time, indicate the trajectories of flows evolving from different initial conditions proportional to the least stable Orr-Sommerfeld mode at this (α, R) . Following an initial transient, flows evolve to a state within a band of quasi-equilibria and reach the steady solution only on times of the order of R.

To date no equilibria have been found in plane Couette flow. There is also no evidence of quasi-equilibria, i.e. all perturbations decay on a time scale short compared to the diffusive scale. In Fig. 8 and Fig. 9 we show the rapid decay of two disturbances. The parameters $(a = 3.0, R = 500)$ of the flow shown in Fig. 8 are in that class predicted to allow equilibria according to amplitude-expansion nonlinear techniques (Davey & Nguyen 1971).

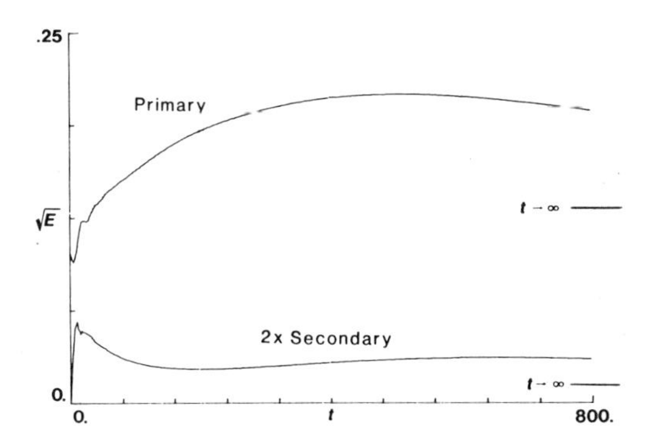

Figure 6. The evolution of a primary disturbance and its harmonic in plane Poiseuille flow at $R = 4000, \alpha = 1.25$ Significant deviations from the steady solution are only eliminated via cross-stream diffusion on a time scale of order R.

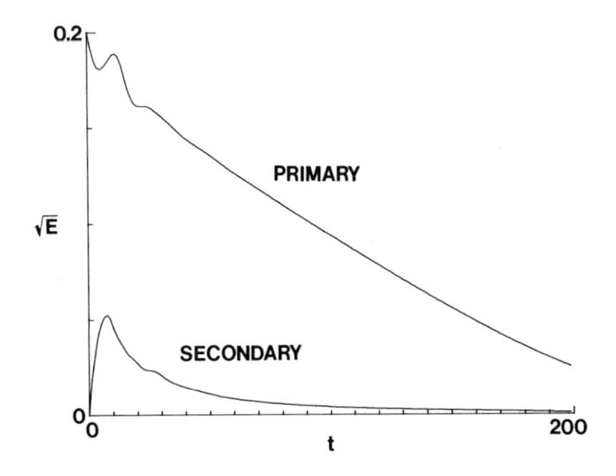

Figure 7. The decay of a disturbance in plane Poiseuille flow at $R = 1500, \alpha = 1.32$. The slow decay at finite - amplitude (four times slower than the linear decay rate) reflects the existence of equilibria at higher Reynolds numbers.

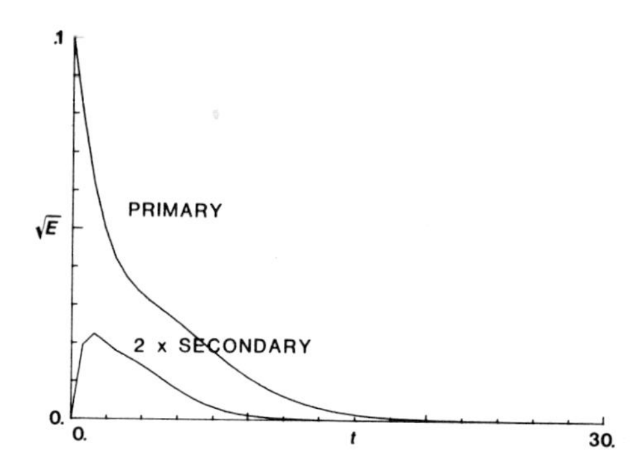

Figure 8. Decay of a finite-amplitude disturbance in plane Couette flow at $\alpha = 3.0, R = 500$. This (α, R) is predicted to be dangerous by amplitude - expansion techniques.

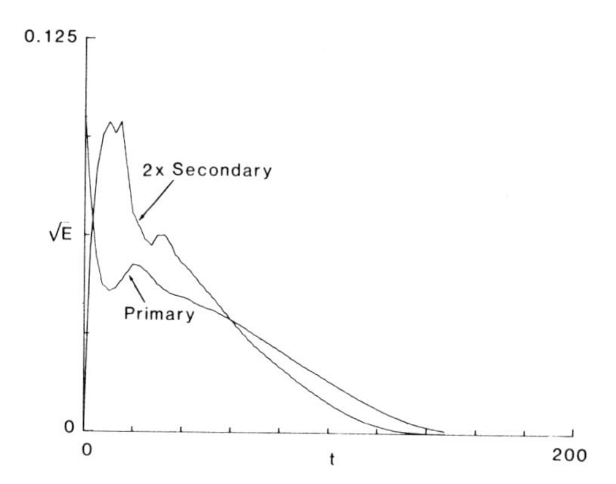

Figure 9. Evolution of a finite-amplitude disturbance in plane Couette flow at $(\alpha = 1.25, R = 4000)$. Note the disturbance decays on a time scale much shorter than the diffusive time scale, R, and we therefore do not expect equilibria nearby in the parameter space.

3 Three-Dimensional Instability

Numerical solution of the Navier-Stokes equations for plane channel flows shows that a very small three - dimensional disturbance superposed on a basic (parallel) flow and a two - dimensional secondary flow undergoes rapid exponential growth. We illustrate this in Figs. 10 and 11 for plane Poiseuille and plane Couette flows, respectively, in which we plot the logarithm of three - dimensional energy, E_{3-D}, vs. time for flows resulting from the initial conditions

$$\mathbf{v}(x, y, z, t = 0) = \overline{U}(z)\hat{\mathbf{x}} + A\mathbf{v}_{2-D}(x, z) + \epsilon\mathbf{v}_{3-D}(x, y, z),$$

where $\overline{U}(z) = (1 - z^2)$ for plane Poiseiulle flow and $\overline{U}(z) = z$ for plane Couette flow. The initial two - dimensional disturbance $\mathbf{v}_{2-D}$ is an Orr-Sommerfeld mode with wave vector $(\alpha, 0)$ and amplitude, A , corresponding to an energy E_{2-D} of 0.04 . The initial three - dimensional disturbance $\epsilon\mathbf{v}_{3-D}$ is an infinitesimal ($\epsilon = 10^{-8}$) Orr-Sommerfeld mode with wave vector $(0, \beta)$.

Observe from Figs. 10 and 11 that this instability singles out a critical Reynolds number of about 1000 in good agreement with transitional Reynolds numbers observed experimentally. The instability is only effective in forcing transition if the three - dimensional perturbation growth rate (and/or initial energy) is sufficiently large *and* the decay rate of the secondary two - dimensional flow is sufficiently small to allow three - dimensional nonlinear effects to develop. Therefore, from the results of Section 2 we would expect three - dimensional threshold energies in plane Couette flow to be greater than in plane Poiseuille flow due to the larger two - dimensional decay rates associated with the former flow. In fact, it will be shown in the next section that for $R > 2900$ in plane Poiseuille flow (i.e. where equilibria do exist) *infinitesimal* three - dimensional perturbations are sufficient to trigger and maintain the instability.

Note the time scale (typically of order 10 for amplitude growth by a factor of 10) on which the three - dimensional perturbation grows is one to two orders of magnitude smaller than the viscous time scales associated with the growth of supercritical (e.g. $R = 10^4$) linear disturbances. This leads us to consider the three - dimensional instability as "inviscid" in the sense that the growth rate σ becomes independent of R as $R \to \infty$ (see Fig. 10).

The time scales relevant to the three - dimensional growth are the mean flow and primary convective scales, $h/\hat{U}$ and $h/\sqrt{E_{2-D}}$, respectively, where $\hat{U}$ characterizes the mean flow velocity. Although it is easily verified that σ scales with $\hat{U}/h$, the dependence on $\sqrt{E_{2-D}}/h$ is more complicated. There does appear to be a region of simple proportionality, however threshold and saturation phenomena at large amplitude are also present. For instance, at $R = 1500$, an initial two - dimensional amplitude of approximately 0.1 - 0.2 gives strong three - dimensional growth, whereas a two - dimensional amplitude of 0.01 results in linear decay of both the two - dimensional and three - dimensional disturbances.

The effect of streamwise wavenumber α reflects the trade-off between three - dimensional growth and two - dimensional decay. (The effect of spanwise wavenumber β is briefly described in Section 4). For instance, in plane Poiseuille flow $\alpha = 1.32$ does not result in the largest three - dimensional growth rate for given two - dimensional energy, although it is close to that wavenumber giving the least stable two - dimensional flow. This is illustrated in Fig. 12. The disturbances followed in Fig. 10 are considered "most dangerous" in the sense that two - dimensional decay appears more significant in determining the low Reynolds number cut-off than three - dimensional perturbation behavior.

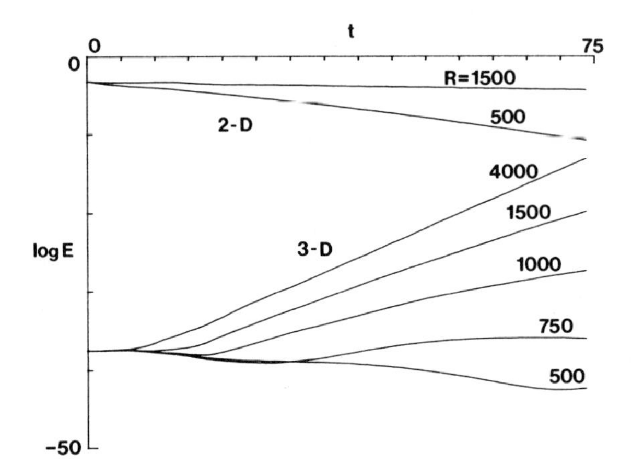

Figure 10 . A plot of the growth of three - dimensional perturbations on finite-amplitude two - dimensional states in plane Poiseuille flow at $(\alpha,\beta) = (1.32, 1.32)$. Here E_{2-D} is the total energy (relative to the basic laminar flow) in wavenumbers of the form $(n\alpha, 0)$, while E_{3-D} is the total energy in wavenumbers $(n\alpha, \beta)$. For $R > 1000$ we obtain growth, whereas at $R = 500$ the three - dimensional perturbations ultimately decay. The growth rate of the three - dimensional disturbance amplitude at $R = 4000$ is about 0.18, and depends only weakly on R as $R \to \infty$.

Varying the spatial resolution of our calculations indicates that the instability reported here is not peculiar to any low-order model but is in fact a property of the Navier-Stokes equations. The large parameter space (e.g. α, R, E_{2-D}) which describes the instability precludes an exhaustive study of the characteristics of the phenomenon, so we restrict ourselves here to demonstrating that the mechanism predicts transition in accordance with experiment. The large growth rates, small threshold energies, and experimentally reasonable dependence on Reynolds number indicate that this instability leads to transition in plane channel flows. The strength of the instability explains why the finite-amplitude two - dimensional states (1) have never been observed.

4 Linear Perturbation Analysis

The existence of equilibria for plane Poiseuille flow allows us to study the mechanism of three - dimensional growth in the relatively simple case in which the the time dependent behaviour of the two - dimensional component need not be considered. In particular, the exponential growth of three - dimensional perturbation energy in Figs. 10 and 11 indicates that a linear mechanism is involved. In this Section we derive and solve the stability equations resulting from linearization of the Navier-Stokes equations around the finite-amplitude (non-parallel) flow (1). (The procedure of finding non-linear two-dimensional states and performing a linear stability analysis about them has been used previously by Clever & Busse 1974 for the case of Benard convection.) In the rest frame (moving with speed c relative to the laboratory frame) the stability problem is separable in time and

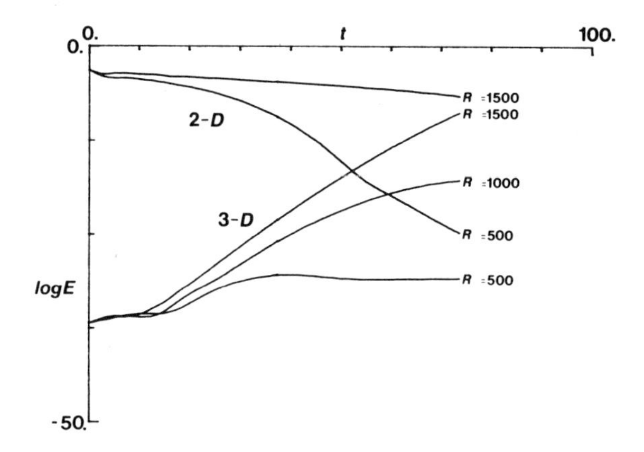

Figure 11. A plot of the growth of three - dimensional perturbations on finite-amplitude two - dimensional states in plane Couette flow at $(\alpha, \beta) = (1.0, 1.0)$. E_{2-D} and E_{3-D} are defined as in Fig. 10 The instability singles out a critical Reynolds number on the order of 1000 in accordance with experiment. The large decay rates of the two - dimensional states in plane Couette flow imply that larger threshold three - dimensional energies are required to force transition in this flow than in plane Poiseuille flow.

periodic in x, and so we assume a solution of the form

$$\mathrm{v}(x, y, z, t) = (1 - z^2)\hat{\mathrm{x}} + \mathrm{F}(x, z) + \epsilon\Re e\{e^{\sigma t} \sum_{m=-1}^{1} \sum_{n=-\infty}^{\infty} \mathrm{u}_{nm}(z) e^{ian x} e^{i\beta m y}\} \tag{8}$$

where σ is a complex frequency and β is the spanwise wavenumber.

Upon inserting (8) into the Navier-Stokes equations and linearizing with respect to ϵ, the following equations result:

$$\begin{aligned}&\{\sigma(D^2 - k^2{}_{n,m}) - \nu(D^2 - k^2{}_{n,m})^2\}w_{n,m} - im\beta D\{(\overline{U} * v_x)_{n,m} + (\overline{W} * v_z)_{n,m}\}\\ &\quad - ina D\{(\overline{U} * u_x)_{n,m} + (u * \overline{U}_x)_{n,m} + (\overline{W} * u_z)_{n,m} + (w * \overline{U}_z)_{n,m}\}\\ &\quad - k^2{}_{n,m}\{(\overline{U} * w_x)_{n,m} + (u * \overline{W}_x)_{n,m} + (\overline{W} * w_z)_{n,m} + (w * \overline{W}_z)_{n,m}\} = 0\end{aligned} \tag{9}$$

$$\begin{aligned}&\{\sigma - \nu(D^2 - k^2{}_{n,m})\}\zeta_{n,m} - ian\{(\overline{U} * v_x)_{n,m} + (\overline{W} * v_z)_{n,m}\}\\ &\quad + im\beta\{(\overline{U} * u_x)_{n,m} + (u * \overline{U}_x)_{n,m} + (\overline{W} * u_z)_{n,m} + (w * \overline{U}_z)_{n,m}\} = 0\end{aligned} \tag{10}$$

$$ianu_{n,m} + im\beta v_{n,m} + Dw_{n,m} = 0 \tag{11}$$

$$im\beta u_{n,m} - ianv_{n,m} = \zeta_{n,m} \tag{12}$$

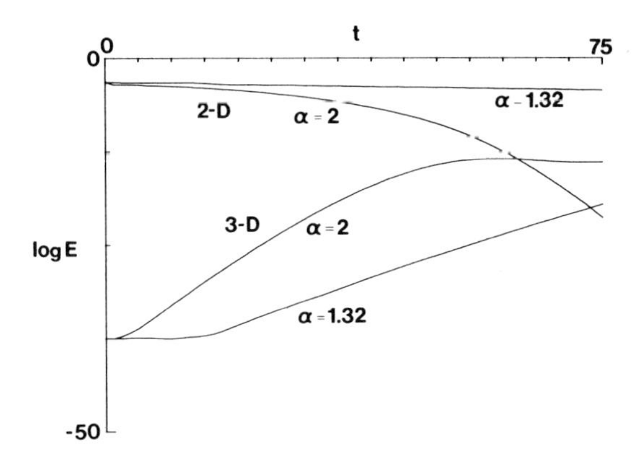

Figure 12. The growth of three - dimensional disturbances on finite–amplitude states in plane Poiseuille flow at $R = 1500$. At $\alpha = 2$ the three - dimensional growth rate is larger than at $\alpha = 1.32$, however strong two - dimensional decay limits the effectiveness of the instability in forcing transition. At large times the three - dimensional instability is turned off. At $\alpha = 1.32$ the relatively steady two - dimensional flow permits persistent growth ultimately resulting in nonlinear three - dimensional interactions.

where $D = \partial/\partial z$, $(\overline{U}, \overline{W})$ is the two - dimensional flow (1) , $k^2{}_{n,m} = \alpha^2 n^2 + \beta^2 m^2$, and the convolution operator $*$ is defined by

$$(h * g)_{n,m} = \sum_{p+q=n} h_{p,m} g_{q,m}.$$

Here u, v and w are the streamwise (x), spanwise (y) , and cross-stream (z) velocities, respectively, and ς is the cross-stream vorticity. Elimination of $\varsigma_{n,m}$ and $v_{n,m}$ in (9) and (10) results in one fourth order equation for $w_{n,m}$ and one second order equation for $u_{n,m}$. At the wall the perturbation velocities must satisfy the no-slip condition

$$w_{n,m}, \quad Dw_{n,m}, \quad u_{n,m} = 0 \qquad (z = \pm 1).$$

At this point we invoke a result from our full numerical simulations of the Navier-Stokes equations in order to reduce the number of unknowns and, hence, simplify the problem. The solutions to the Navier-Stokes equations indicate that the three - dimensional perturbation travels at the same speed as the two - dimensional finite-amplitude wave, F , and so $\mathfrak{Im}\ \sigma = 0$ in (8). (This assumption is validated by the solution of the eigenvalue problem (9) - (12)). The spatial dependence of the three - dimensional perturbation in (8) must therefore be real, or

$$\mathbf{u}_{n,m} = \mathbf{u}^{\dagger}{}_{-n,-m}.$$

In addition we note the following symmetries consistent with (9) - (12),

$$\{u_{n,m}(z), v_{n,m}(z), w_{n,m}(z)\} = \pm(-1)^{n+1}\{u_{n,m}(-z), v_{n,m}(-z), -w_{n,m}(-z)\} \qquad (13)$$

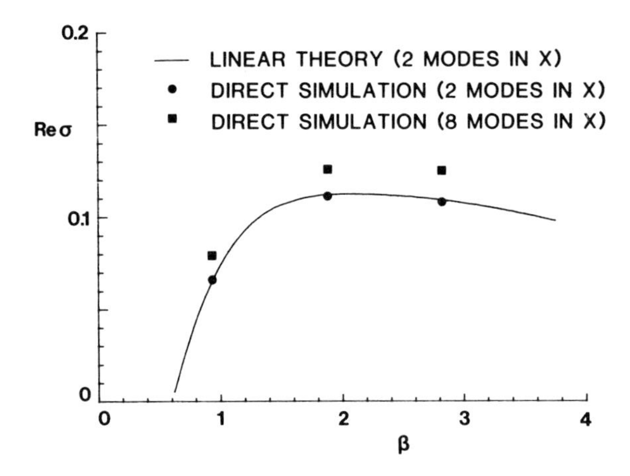

Figure 13 . A plot of the growth rate of three - dimensional perturbations, σ , as a function of β at $R = 4000, a = 1.25$. Note the good agreement between the linear calculation and the full simulation. Increasing the number of nodes in the x - direction increases the growth rate, but the error in the $N = 2$ – mode model is not large.

and

$$\{u_{n,m}(z), v_{n,m}(z), w_{n,m}(z)\} = \pm\{u_{n,-m}(z), -v_{n,-m}(z), w_{n,-m}(z)\} \tag{14}$$

In this section we restrict ourselves to the upper signs in (13) and (14) exclusively. The three symmetries above allow us to solve for only the modes $m = 1, n \geq 0$ on half the channel.

The numerical procedure used to solve the linear eigenvalue problem (9) - (12) is similar to that used to solve the nonlinear equations (3). A Galerkin procedure is used in x (truncated at $n = N - 1$) and Chebyshev collocation is used in z. The dynamical equation at the wall is dropped in favor of one boundary condition . Additional boundary conditions (e.g. $Dw_{n,m} = 0$) result in additional equations and corresponding τ – factors. If K is the number of collocation points in the half-channel, the number of (real) unknowns is $K(4N - 2) + N$. The algebraic eigenvalue problem is solved locally with a Newton iteration or globally with the QR algorithm. When a local algorithm is used, an arbitrary normalization is required so the matrix is of rank $K(4N - 2) + N + 2$.

The (real) growth rate predicted by the linear problem (9) - (12), σ , is plotted vs. β at $R = 4000, a = 1.25$ with $N = 2, K = 17$ in Fig. 13. The good agreement between these growth rates and those obtained from the full simulations indicates that the linear mechanism isolated here is indeed responsible for the growth of three - dimensional perturbations described in the previous section. The effect of increasing N (which is more easily done in the full simulation than in the linear problem) is to *increase* σ , however the error in the $N = 2$ – mode model is not large. Increasing K indicates that the solution has converged in z. From Fig. 12 we also see that the growth rate is fairly insensitive to β once a certain "threshold" three-dimensionality is achieved.

Although the linear analysis presented here strictly applies only when equilibria exist, separation of time scales allows it to be used where quasi - equilibria exist as well by freezing the two - dimensional motion.

5 Discussion

The "inviscid" nature of the instability described here might suggest an explanation based on the instability of instantaneously inflectional velocity profiles created by the two - dimensional eddy motions. However, such an explanation could not adequately describe the most distinctive characteristic of the instability, namely its three-dimensionality. Indeed, the classical results of Squire, Fjortoft, and Rayleigh show that three - dimensional inflectional inviscid instability implies two - dimensional inflectional inviscid instability, which is not true here.

The complexity of the three - dimensional instability can best be described by vorticity dynamics, i.e. the stretching and tilting of vortex lines. In particular, we can write the equation for the perturbation vorticity as

$$\frac{\partial \omega_1}{\partial t} + \{(\mathbf{v}_0 \cdot \nabla)\omega_1 - (\omega_1 \cdot \nabla)\mathbf{v}_0\} + \{(\mathbf{v}_1 \cdot \nabla)\omega_0 - (\omega_0 \cdot \nabla)\mathbf{v}_1\} = 0, \tag{15}$$

where subscript 0 refers to a two - dimensional equilibrium (independent of y) and subscript 1 denotes the three - dimensional perturbation. We neglect the effect of viscosity and assume periodic boundary conditions. Note that conserved "inviscid" integrals of motion will in fact decay in the presence of finite dissipation.

The first bracketed term in (15) represents the action of the two - dimensional flow on the three - dimensional perturbation (primarily stretching), while the second term represents the action of the perturbation on the two - dimensional flow (primarily tilting). We now show (within the periodic restriction outlined above) that any mechanism proposed to explain the three - dimensional instability must include both these effects. Thus, a simple explanation of the phenomenon based on stretching of perturbation vortex filaments in the plane of the two - dimensional flow, i.e. a "vortex" dynamo, cannot be valid, just as two - dimensional magnetic dynamos do not exist (Moffatt 1978).

Including only the effects of tilting, (15) becomes

$$\frac{\partial(\nabla \times \mathbf{v}_1)}{\partial t} = (\omega_0 \cdot \nabla)\mathbf{v}_1 - (\mathbf{v}_1 \cdot \nabla)\omega_0 = \nabla \times (\mathbf{v}_1 \times \omega_0)$$

where $\omega_1 = \nabla \times \mathbf{v}_1$. Upon integrating out the curl and forming the energy integral, we have

$$\frac{1}{2}\frac{\partial}{\partial t}\int (\mathbf{v}_1 \cdot \mathbf{v}_1)\, d\mathbf{x} = \int \mathbf{v}_1 \cdot (\mathbf{v}_1 \times \omega_0)\, d\mathbf{x} - \int (\mathbf{v}_1 \cdot \nabla\phi)\, d\mathbf{x}$$

where ϕ is a potential. The integrand of the first term on the right-hand side vanishes identically. The second term can be converted to a surface integral which vanishes due to periodicity. Thus we find that tilting alone cannot result in rapid growth.

On the other hand, including only the stretching effects, (15) becomes

$$\frac{\partial \omega_1}{\partial t} = (\omega_1 \cdot \nabla)\mathbf{v}_0 - (\mathbf{v}_0 \cdot \nabla)\omega_1 = \nabla \times (\mathbf{v}_0 \times \omega_1).$$

It follows that the convective derivative of $\omega_1 \cdot \hat{\mathbf{y}}$ vanishes so that we can restrict attention to a two - dimensional solenoidal vorticity field, $\omega_1 = \nabla \times \psi\hat{\mathbf{y}}$. Upon integrating out the curl we get

$$\frac{\partial\psi(x,y,z,t)}{\partial t} + (\mathbf{v}_0(x,z) \cdot \nabla)\,\psi(x,y,z,t) = -\frac{\partial\phi(y,t)}{\partial t} \tag{16}$$

Including the potential ϕ in the convective derivative (noting $\mathbf{v}_0 \cdot \hat{\mathbf{y}} = 0$), we obtain

$$\frac{\partial(\psi + \phi)}{\partial t} + (\mathbf{v}_0 \cdot \nabla)\,(\psi + \phi) = 0.$$

Thus it follows that $\omega_1 \rightarrow 0$ and $\frac{1}{2}\int(\mathbf{v}_1 \cdot \mathbf{v}_1)\,d\mathbf{x} \rightarrow 0$ (since the only potential flow in a periodic domain is $\mathbf{v} = 0$). We conclude that the physics of the present instability involves a delicate balance between vortex tilting and vortex strething.

This work was supported by the Office of Naval Research under Contracts No. N00014-77-C-0138 and No. N00014-79-C-0478. Development of the stability codes was supported by NASA Langley Research Center under Contract No. NAS1-15894. The computations were performed at the Computing Facility of the National Center for Atmospheric Research, which is supported by the National Science Foundation.

Appendix: Numerical Methods

As the treatment of the nonlinear terms in the Navier-Stokes equations is the same using both splitting methods (Orszag & Kells 1980) and full- step techniques, we will not discuss them here. The differences between the two methods is best illustrated in the context of the Stokes equations

$$\frac{\partial\mathbf{v}}{\partial t} = -\nabla\Pi + \nu\nabla^2\mathbf{v}$$

$$\nabla \cdot \mathbf{v} = 0 \tag{A1}$$

$$\mathbf{v} = 0, \qquad z = \pm 1.$$

The splitting method factors these equations into a pressure operator (to impose incompressibility) and a viscous operator. As these operators do not commute when no-slip boundary conditions are imposed, one must incur a first-order error in time (though in practice this is not found to be restrictive for shear flows). Using the splitting technique each time step requires the solution of four Poisson equations.

In the full-step method we reduce the Stokes equations to a fourth order equation for the cross-stream velocity, w . Using the (second - order) Crank - Nicolson method to advance the solution in time, the equation for w at time t is then given by

$$\{D^2 - k^2\}w = \varsigma$$

$$\{D^2 - (k^2 + \frac{2}{\nu\Delta t})\}\varsigma = f \tag{A2}$$

$$w = Dw = 0 \qquad z = \pm 1$$

where $D = \partial/\partial z$, k^2 represents a Fourier wavenumber, and Δt is the time-step. Here f represents a forcing term (or the results of the nonlinear step). This coupled system is solved using a Chebyshev-tau / Green's function technique, in which we write the solution as

$$\varsigma_m = \sum_{i=1}^{5} \tau^i \varsigma_m^i$$

$$w_m = \sum_{i=1}^{5} \tau^i w_m^i$$

where, in general, h_m is the coefficient of the Chebyshev polynomial of degree m in a Pth – order spectral expansion of $h(z)$. The equations for the w^i and the ς^i are then given by

$$\{D^2 - k^2\} w_m^i = \varsigma_m^i \qquad 0 \leq m \leq P-2$$

$$\sum_{m=0}^{P} w_m^i (\pm 1)^m = 0,$$

and

$$\{D^2 - (k^2 + \frac{2}{\nu \Delta t})\} \varsigma_m^i = e_m^i \qquad 0 \leq m \leq P-2$$

$$\sum_{m=0}^{P} \varsigma_m^i (\pm 1)^m = e_n^i \qquad P-1 \leq n \leq P$$

respectively. The τ^i $(i = 1, \ldots, 4)$ are determined by the four equations

$$\{D^2 - k^2\} w_m = \varsigma_m \qquad P-1 \leq m \leq P$$

$$\sum_{p=0}^{P} (\pm 1)^p p^2 w_p = 0,$$

and $\tau^5 = 1$. The e_m^i are given by

$$e_m^1 = \delta_{m,P-1}$$
$$e_m^2 = \delta_{m,P}$$
$$e_m^3 = D(T_P)_m$$
$$e_m^4 = D(T_{P-1})_m$$
$$e_m^5 = f_m,$$

where T_P is the Chebyshev polynomial of degree P, and δ is the Kronecker - delta function. Thus w_m^5, ς_m^5 is a particular solution of (A2) while w_m^i, ς_m^i $(i \leq 4)$ are used to enforce the boundary conditions.

The solution of the fourth order equation is thus reduced to the solution of four Poisson equations per timestep per Fourier mode. In order to complete the timestep an additional Poisson equation (for the cross-stream vorticity) must also be solved. Each of the Poisson equations can be represented as an essentially tri-diagonal diagonally-dominant system (Gottlieb & Orszag 1977) which can be efficiently inverted. A detailed comparison of the accuracy of these methods will be given elsewhere.

References

Benney, D. J. & Lin, C. C. 1960 On the secondary motion induced by oscillations in a shear flow. *Phys. Fluids* **3**, 656.

Clever, R. M. & Busse, F. H. 1974 Transition to time-dependent convection. *J. Fluid Mech.* **65**, 625.

Craik, A. D. D. 1980 Nonlinear evolution and breakdown in unstable boundary layers. *J. Fluid Mech.* **99**, 247.

Davey, A. & Nguyen, H. P. F. 1971 Finite-amplitude stability of pipe flow. *J. Fluid Mech.* **45**, 701.

Davies, S. J. & White, C. M. 1928 An experimental study of the flow of water in pipes of rectangular section. *Proc. Roy. Soc.* A **119**, 92.

Gottlieb, D. & Orszag, S. A. 1977 *Numerical Analysis of Spectral Methods: Theory and Applications.* NSF-CBMS Monograph no. 26, Soc. Ind. App. Math., Philadelphia.

Gustavsson, L. H. & Hultgren, L. S. 1980 A resonance mechanism in plane Couette flow. *J. Fluid Mech.* **98**, 149.

Herbert, Th. 1976 Periodic secondary motions in a plane channel. In *Proc. 5th Int. Conf. on Numerical Methods in Fluid Dynamics* (ed. A. I. van de Vooren and P.J. Zandbergen),p. 235. Springer.

Herbert, Th. 1977 Finite amplitude stability of plane parallel flows. In *Laminar-Turbulent Transition, AGARD Conf. Proc.* no. 224,p. 3-1.

Herbert, Th. 1980 Nonlinear stability of parallel flows by high-order amplitude expansions. *AIAA Journal* **18**, 243.

Herbert, Th. & Morkovin, M. V. 1980 Dialogue on bridging some gaps in stability and transition research. In *Laminar-Turbulent Transition, IUTAM Conf. Proc.* (ed. R. Eppler and H. Fasel),p. 47. Springer.

Kao, T. W. & Park, C. 1970 Experimental investigation of the stability of channel flows. Part 1. Flow of a single liquid in a rectangular channel. *J. Fluid Mech.* **43**, 145.

Lentini, M. & Keller, H. 1980 The von Karman swirling flows. *SIAM J. Applied Math.* **38**, 52.

Meksyn, M. & Stuart, J. T. 1951 Stability of viscous motion between parallel planes for finite disturbances. *Proc. Roy. Soc.* A **208**, 517.

Moffatt, H. K. 1978 *Magnetic Field Generation in Electrically Conducting Fluids.* Cambridge University Press, Cambridge,p.121.

Nishioka, M., Iida, S. & Ichikawa, Y. 1975 An experimental investigation of the stability of plane Poiseuille flow. *J. Fluid Mech.* **72**, 731.

Orszag, S. A. & Kells, L. C. 1980 Transition to turbulence in plane Poiseuille and plane Couette flow. *J. Fluid Mech.* **96**, 159.

Orszag, S. A. & Patera, A. T. 1980 Subcritical transition to turbulence in plane channel flows. *Phys. Rev. Lett.* **45**, 989.

Patel, V. & Head, M. R. 1969 Some observations on skin friction and velocity profiles in fully developed pipe and channel flows. *J. Fluid Mech.* **38**, 181.

Patera, A. T. & Orszag, S. A. 1980a Transition and turbulence in plane channel flows. In *Proc. 7th Int. Conf. on Numerical Methods in Fluid Dynamics* (ed. R. W. MacCormack and W. C. Reynolds), Springer.

Patera, A. T. & Orszag, S. A. 1980b Finite amplitude stability of axisymmetric pipe flow. To be published.

Reichardt, H. 1959 Gesetzmassigkeiten der geradlinigen turbulenten Couettestromung. *Mitt. Max-Planck-Inst. StromForsch.* **22**, Gottingen.

Stuart, J. T. 1960 On the non-linear mechanics of wave disturbances in stable and unstable parallel flows. Part 1. The basic behaviour in plane Poiseuille flow. *J. Fluid Mech.* **9**, 353.

Zahn, J.-P., Toomre, J., Spiegel, E. A. & Gough, D. O. 1974 Nonlinear cellular motions in Poiseuille channel flow. *J. Fluid Mech.* **64**, 319.

Department of Mathematics
Massachusetts Institute of Technology
Cambridge, Massachusetts 02139

Remark on Engineering Aspects of Transition

E. Reshotko

In discussing transition, one must distinguish between the scientific and the engineering aspects of transition. The scientific aspects of transition encompass the complete process. The engineering aspects of transition are those of concern for the prediction and/or control of the transition process.

In an environment where initial disturbance levels are small, the transition Reynolds number of a boundary layer, for example, is very much dependent upon the nature and spectrum of the disturbance environment, the signatures in the boundary layer of these disturbances and their excitation of the normal modes (referred to in the literature as "receptivity"),and finally the linear amplification of the growing normal modes. The non-linear processes which follow serve to bring transition quickly to completion once they set in. There is ample documentation that the factors that affect linear amplification are the primary factors that determine the magnitude of the transition Reynolds number. This is simply because the linear amplification step is the slowest of the multiple steps in the transition process.

There are some important mathematical questions arising in conjunction with the initiation of transition and linear amplification. Fortunately, these questions all relate to the regimes where linear disturbance equations suffice.

ISBN 0-12-493240-1

First, what are the complete set of normal modes, both discrete and continuum, and can the completeness be demonstrated? Second, can any of the common external stimuli (free-stream turbulence, acoustic excitations, roughness, etc.) be represented in terms of normal modes? Third, is there any general way of handling the receptivity question for a growing (non-parallel) boundary layer?

Another question relating to the control of transition is: How far can the transition process proceed and yet be reversed? Professor Gary Brown may perhaps answer this question when the full results of his water tunnel experiments, described at this Symposium, become available.

A most important and practical question is that of boundary layer transition in the presence of large (non-linear) initial disturbances. There is some feeling that the boundary layer is different in the respect from plane Poiseuille flow because the latter is a fully parallel flow whereas for the boundary layer, the region of importance is the growing (non-parallel) region near the leading edge. The boundary layer problem is mathematically much more difficult than the plane Poiseuille problem.

We have heard much at the Symposium about the development and consequences of the non-linear portions of the transition process including wave-interactions, spot formation and the development of turbulent structures. The processes at the beginning of transition are no less important and a proper mathematical and physical understanding of their details would be most welcome.

Department of Mechanical and
Aerospace Engineering
Case Western Reserve University
Cleveland, Ohio 44106

Vortex Interactions and Coherent Structures in Turbulence

P. G. Saffman

1. The Significance of Vorticity.

The importance of vorticity for the understanding and description of the turbulent flow or high Reynolds number motion of a uniform, incompressible fluid cannot be overemphasized. From the definition of vorticity $\underset{\sim}{\omega}(\underset{\sim}{x},t)$ as the curl of the velocity field $\underset{\sim}{u}(\underset{\sim}{x},t)$,

$$\underset{\sim}{\omega} = \text{curl } \underset{\sim}{u}, \tag{1.1}$$

it follows using standard vector identities and the incompressibility condition $\text{div } \underset{\sim}{u} = 0$, that

$$\nabla^2 \underset{\sim}{u} = - \text{curl } \underset{\sim}{\omega} \ . \tag{1.2}$$

Hence the vorticity determines the velocity field through Poisson's equation. The arbitrary irrotational, harmonic flow field $\nabla\phi$, where $\nabla^2\phi = 0$, which can be added to any solution of (1.2), is in general specified uniquely by boundary or periodicity conditions on $\underset{\sim}{u}$. Now as a consequence of the dynamics (expressed by the Helmholtz laws which state that vorticity is convected with the fluid and can neither be created nor destroyed provided that the effects of viscosity are negligible, the flow is free of singularities, and non-conservative forces are absent), it follows that if the vorticity is originally confined to some finite region, then it will always be confined to a finite region and the distribution of vorticity may be described much more

ISBN 0-12-493240-1

economically than that of the velocity field. Moreover, in addition to the usual Eulerian description, the dynamics of vorticity lends itself to a Lagragian description by the equation

$$\frac{d\underset{\sim}{\omega}}{dt} = (\underset{\sim}{\omega}\cdot\nabla)\underset{\sim}{u} + \nu\nabla^2\underset{\sim}{\omega} = \underset{\sim}{\omega}\cdot\underset{\approx}{e} + \nu\nabla^2\underset{\sim}{\omega} \tag{1.3}$$

where $\underset{\approx}{e} = \frac{1}{2}(u_{i,j} + u_{j,i})$ is the rate of strain tensor and does not depend directly on the local vorticity, and ν is the kinematic viscosity. The terms describe, respectively, the rate of change of vorticity following the fluid, the amplification due to the irrotational stretching of vortex lines, and the viscous diffusion which is generally negligible when ν is small except near boundaries or shear layers. The viscous term can be handled in the Lagrangian framework by replacing it with a random walk contribution to the convection velocity, but there are some restrictions; see Milinazzo and Saffman [10]. Note also that although the region of vorticity stays finite, it may become very distorted and the Lagrangian approach may be consequently inferior to the Eulerian description.

Another reason for the significance of vorticity is that the rate of energy dissipation per unit volume may be written

$$\frac{d}{dt}\frac{1}{2}\underset{\sim}{u}^2 = -\nu\underset{\sim}{\omega}^2 + \text{divergence terms}, \tag{1.4}$$

so the energy dissipation is large in the regions of high vorticity.

It can be claimed that the problems of turbulence and transition are problems of the mechanics of vorticity, which is the fundamental dynamical quantity and constitutes the "sinews and muscles of fluid motion." One of the best definitions of turbulence is that it is a field of random or chaotic vorticity. This distinguishes turbulence from noise produced by random motion of boundaries.

During the past 45 years, much effort has been spent trying to determine the statistical distribution and in particular the spectra of the vorticity distribution in turbulence. However, the most exciting recent development is the growing belief, suggested by modern experimental investigations, that the vorticity fluctuations are not quite so random or disorganized or incoherent as was commonly thought. The vorticity is perhaps collected into coherent structures or organized eddies, and it is now proposed that turbulence should be modelled or described as the creation, evolution, interaction and decay of these structures. Turbulence is then thought of as the random superposition of organized, laminar, deterministic vortices, whose life history and relationships constitutes the turbulent flow. A significant theoretical effort is underway, attempting to describe, understand and predict the properties of turbulent flow by the study of vortex dynamics. The present article describes the results of one of the avenues being explored, namely the existence and stability of steady inviscid motions of vortices of finite size. Incidentally, it should be noted that the approach is not new, as nearly 40 years ago Synge and Lin represented turbulence as a superposition of Hill spherical vortices and Townsend and others used similar ideas. For a recent review, see [18].

2. Atomic and Molecular Representations.

Unfortunately, even laminar motions of fluids endowed with vorticity are far from simple and pose problems of great mathematical and numerical complexity. The study of vortex dynamics is a challenging subject in its own right, apart from the applications to transition phenomena and turbulent flow. The questions that are asked are concerned with evolution, equilibrium and stability. Given an initial distribution of vorticity, we wish to know how it moves, whether there are any equilibrium states or limit cycles and if they are stable, and so on. To add to the difficulty, it has not been proved that the general initial value problem in three-dimensions is mathematically well posed and there exist speculations that it is not for inviscid fluids. For a summary of results of the stability problem for ideal flow, see Arnold [1].

A useful and popular method of analysing vortex flows, especially for small viscosity and concentrated vorticity, replaces the actual continuous distribution of vorticity by a finite sum over a number of discrete vortices, defined as a volume (area in two-dimensional flow) of rotational fluid surrounded by irrotational fluid. For want of better words, we call this representation 'atomic' or 'molecular' according as the discrete vortices are infinitesimal or finite. In the atomic representation, the structure of the vortices is supposed given and their motion is described by the Lagrangian evolution equation (1.3). In the molecular representation, the deformation and structure of the vortex are at least as important as the motion of the vortex as a whole.

The atomic approach is particularly simple for two-dimensional flows when the vortices are usually either points or circles of small radius. The techniques of complex variable can be used for analytical studies and the evolution equations reduce to a set of ordinary differential equations for numerical work. The literature is vast; analytical work dates back more than 100 years and Rosenhead's numerical studies were carried out in 1931. See [19] for a recent review. For three-dimensional flows, things are not quite so easy, but progress is possible. The principal method used so far represents the vorticity as a collection of vortex filaments (i.e. vortex tubes of small, assumed circular, cross section) and the velocity is calculated from either the Biot-Savart law of induction with a cut-off proportional to the filament radius to give a finite (but logarithmic in the radius) value to the self induced velocity, or by redistributing the vorticity on to mesh points of a grid and using a Poisson solver. Leonard [8] has reviewed this work and describes encouraging results on modelling three-dimensional boundary layer perturbations.

A new approach, but still very much in the initial stages of development, employs the idea of a vorticity pole or 'vorton'. As in the two-dimensional case, the vorticity is expressed as a sum of point singularities

$$\underset{\sim}{\omega}(\underset{\sim}{x},t) = \sum_i \underset{\sim}{\kappa}_i(t)\ \delta(\underset{\sim}{x}-\underset{\sim}{x}_i(t)) \qquad (2.1)$$

where the i-th vorton has strength $\underset{\sim}{\kappa}_i$ at position $\underset{\sim}{x}_i$. The representation violates $\operatorname{div} \underset{\sim}{\omega} = 0$ locally but, like the use of magnetic poles for Maxwell's equations, this does not necessarily invalidate the results. The place where care is needed is the evaluation of the velocity field. A consistent approach for simply connected regions is to calculate the velocity potential from

$$\nabla^2 \underset{\sim}{A} = -\underset{\sim}{\omega} \ , \tag{2.2}$$

with boundary conditions $\underset{\sim}{n} \wedge \underset{\sim}{A} = 0$, $\partial A_n/\partial n = 0$, which ensures $\operatorname{div} \underset{\sim}{A} = 0$. Then $\underset{\sim}{u} = \operatorname{curl} \underset{\sim}{A}$ and the evolution equation becomes

$$\frac{d\underset{\sim}{x}_i}{dt} = \underset{\sim}{u}(\underset{\sim}{x}_i) \ , \quad \frac{d\underset{\sim}{\kappa}_i}{dt} = \underset{\sim}{\kappa}_i \cdot \nabla \underset{\sim}{u}(\underset{\sim}{x}_i) \ . \tag{2.3}$$

Unfortunately, no results are yet available.

The molecular representation has become of topical importance because of the experimental observations of coherent, quasi-permanent vortices and the renewed interest in the existence of steady configurations and their stability and origin. The evolution of finite vortices can of course be studied using an atomic representation, e.g. [3]. But direct methods are preferable and recent developments in both computer hardware and numerical techniques have made these available, e.g. the water bag model [22] can now be implemented for the evolution of finite vortices.

Steady states can be calculated in some special cases by analytical means; Kirchhoff-Helmholtz free streamline theory can be applied to stagnant cored vortices and there are exact closed form solutions for elliptical uniform vortices, but it is in general necessary to use computational methods for quantitative results. The essence of the approach in two dimensions is to note that the stream function $\psi(x,y)$, $u = \psi_y$, $v = -\psi_x$, produced by a vortex of area A is

$$\psi = -\frac{1}{4\pi} \iint_A \omega(x',y') \log\{(x-x')^2 + (y-y')^2\}\, dx'\, dy'. \tag{2.4}$$

The motion is steady in a frame translating with velocity U,V and rotating with angular velocity Ω if

$$\Psi = \Sigma\,\psi - Uy + Vx + \Omega(x^2+y^2) = \text{const.} \tag{2.5}$$

on the boundary of each vortex, where $\Sigma\psi$ is over the vortices and may include externally imposed irrotational flow fields. The vorticity ω is not completely arbitrary since the Helmholtz laws require

$$\omega = F(\Psi), \tag{2.6}$$

but F is at our disposal. Given this function, the mathematical problem is to find the shapes of the vortices such that (2.4), (2.5) and (2.6) are consistent. The question of the best F, such that the molecular representation gives a good description of the actual flow field, is an unsolved problem. Most work takes either ω = const. i.e. uniform vortices, or $\omega = 0$ except on the boundaries where it is singular corresponding to a vortex sheet surrounding a stagnant core. These choices offer the weighty mathematical advantage of reducing the area integral in (2.4) to a line integral. Physically, the second choice is relevant to the vortices produced by cavitation in liquids and the former has some justification in the Prandtl-Batchelor theorem for steady flows which states that $\omega \to$ const. as $\nu \to 0$ inside a region bounded by closed streamlines. The resulting one-dimensional integral equation is solved by iteration or Newton's method and families of solutions are obtained.

It should, however, be kept in mind that there are strong reasons for believing that ω is not uniform and is strongly peaked in many cases [19], so molecular representations employing uniform vortices may not always be appropriate.

The steady flows are associated with the critical points of the kinetic energy functional for isovorticial [1] variations which conserve linear and angular impulse, as noted by Lord Kelvin. In two dimensions, this takes the form

$$\delta \int \omega \, \psi \, dx \, dy = 0, \tag{2.7}$$

subject to

$$\delta \int \omega y \, dx \, dy = 0, \qquad \delta \int \omega x \, dx \, dy = 0, \qquad \delta \int \omega(x^2+y^2) dx \, dy = 0, \tag{2.8}$$

for variations $\delta \underset{\sim}{s}$ which conserve circulation and volume, i.e.

$$\delta\omega = -\delta \underset{\sim}{s}\cdot\nabla\omega, \qquad \text{div}\ \delta \underset{\sim}{s} = 0 . \tag{2.9}$$

As pointed out by Kelvin, if the energy is an absolute maximum or minimum (and certain other conditions are satisfied [1]) then the flow is stable. This argument has proved useful in obtaining qualitative criteria for stability. Quantitative investigations of two-dimensional stability are rather difficult and progress is slow.

Certain questions, such as existence and uniqueness proofs for large vortices and the stability to general three-dimensional disturbances still seem outside present theoretical and computational capabilities.

3. The Isolated Two-dimensional Vortex.

The simplest solution is the uniform rectilinear vortex with circular boundary and streamlines which are circles around the center. This is, however, not a unique solution for an isolated uniform vortex of strength Γ, area A, and vorticity $\omega_o = \Gamma/A$. Kirchhoff [7, p.232] showed that an elliptical vortex of arbitrary eccentricity is a steady solution in a reference frame rotating with angular velocity $ab\,\omega_o/(a+b)^2$, where a and b are the axes. Further, it was recently noted by Deem and Zabusky [5] that the Kirchhoff vortex is just the $m=2$ case of an infinite number of families of non-circular shapes with m-fold symmetry, which can be regarded as bifurcations from the circle. This is because a perturbation proportional to $e^{im\theta}$ has time dependence $e^{-i\sigma t}$ where $\sigma = \frac{1}{2}\omega_o(m-\text{sgn } m)$. Hence the circular vortex bifurcates into families of steadily rotating vortices with m-fold symmetry initially rotating with angular velocity $\frac{1}{2}\omega_o(1-1/|m|)$. As the branch is followed, the vortex becomes more deformed and the angular velocity has to be calculated as part of the solution. Deem and Zabusky computed triangular $(m=3)$ and square $(m=4)$ rotating vortices. The stability of these shapes remains an open question.

These results should make us aware that the common assumption, made tacitly without exception, that widely separated vortices are almost circular is not necessarily valid. Recently, Moore [11] has examined the unsteady motion of thin vortex rings with elliptical cores.

4. Vortices in Uniform Strain.

Another, rather more important, generalization of the Kirchhoff vortex was found by Moore and Saffman [12], which can be regarded as the interaction between a vortex and others far away. They examined the possibility of an elliptical vortex, containing uniform vorticity ω, being at rest in a uniform strain and rotation. It was found that steady solutions exist with the ellipse axes aligned at 45° to the principal axes of strain. The vortex bounded by the ellipse

$$\frac{x^2}{a^2} + \frac{y^2}{b^2} = 1, \qquad r = a/b > 1 \tag{4.1}$$

is in equilibrium in the straining field asymptotic at infinity to

$$\psi \sim \frac{1}{2}\varepsilon\,(x^2-y^2) - \frac{1}{2}\gamma\,(x^2+y^2), \tag{4.2}$$

provided

$$\varepsilon\frac{(r+1)}{r-1} - \gamma = \frac{\omega r}{r^2+1} . \tag{4.3}$$

An interesting feature of this result is that for given values of ε, γ and ω, there may be none or more than one solution.

The quantity γ can be interpreted as solid body rotation of the whole system with angular velocity $-\gamma$, the vorticity inside the ellipse then being $\omega - 2\gamma$. Putting $\varepsilon = 0$, and replacing ω by $\omega_o + 2\gamma$, gives $-\gamma = \omega_o r/(r+1)^2$ which is the Kirchhoff result.

When $\gamma = 0$, we have the case of a uniform vortex in an irrotational strain. It is easily shown that there are two elliptical solutions if $\varepsilon/\omega < 0.15$ and none otherwise. When two exist, one has $r < 2.9$ and the other has $r > 2.9$.

This result demonstrates the possibility of vortex fission, i.e. the disintegration of a finite vortex when exposed to a too large rate of strain; a result which may have considerable significance for the understanding of the disintegration or amalgamation of coherent structures.

It is possible to analyse the stability of these elliptical vortices to two-dimensional infinitesimal perturbations [12]. A normal mode with azimuthal wave number m has frequency σ where

$$4\,\frac{\sigma^2}{\omega^2} = \left(\frac{2mr}{r^2+1} - 1 + \frac{2\gamma}{\omega}\right)^2 - \left(1 - \frac{2\gamma}{\omega}\right)^2 \left(\frac{r-1}{r+1}\right)^{2m} . \quad (4.4)$$

It is easily shown that of the two vortices that exist when $\gamma = 0$, the less distorted is stable while the more distorted is unstable. Since neutral disturbances correspond to an exchange of stability according to (4.4), it follows further that there exist an infinite number of bifurcations from the elliptical solutions. In particular, the Kirchhoff solution bifurcates when $r = 3$, which is the value found by Love [9] for instability, into non-elliptical solutions with the same 2-fold symmetry. The stability of these shapes remains to be elucidated, but it is a reasonable guess that they are all unstable.

There has been some work done on the three-dimensional stability of these vortices. The stability to long wavelength disturbances was discussed in [12], but the analysis proved difficult and it was not possible to extract more than could be deduced from an atomic approach, i.e. regarding the vortex as a filament and using the Biot-Savart law with a suitable cut-off. Moore and Saffman [13] examined the stability when $\varepsilon/\omega \ll 1$ for vortices of arbitrary structure and demonstrated the existence of a weak, parametric instability with wavelength $\approx$ core radius.

5. Vortex Pairs.

The equilibrium shapes of a uniformly translating pair of equal and opposite uniform vortices has been investigated systematically by Pierrehumbert [15]. Saffman and Szeto [20] have solved the corotating case of two like signed vortices.

Consider first the translating vortex pair; a flow field which is relevant to the Jumbo jet trailing vortex problem and any flow field employing horse shoe vortices as a model. The solutions behave smoothly as A/ℓ^2 increases, where A is the area of each vortex and ℓ the distance between their centroids, with the vortices gradually becoming more deformed and squashed as they deviate from the circular shape for $A/\ell^2 << 1$. Pierrehumbert claims that the solutions tend continuously to a limit at $A/\ell^2 = 59.2$, in which the vortices touch along a common axis and each resembles half an ellipse, but the numerics is not sufficiently precise to put this claim beyond reasonable doubt and there is no known theoretical reason why solutions should not exist for arbitrarily large A/ℓ^2 with the vortices extended indefinitely along the direction of motion, as is the case for the analytic solution for hollow vortices. There is a challenging existence problem here. But in any case, there is no trace of non-uniqueness or symmetrical bifurcation.

For the rotating vortex pair, a smooth solution branch exists as A/ℓ^2 increases from zero, has a limit point at $A/\ell^2 = 5.00$, and terminates at $A/\ell^2 = 4.93$ where the vortices touch at a cusp. It is expected that the solution could then be continued into the single rotating vortex with two-fold symmetry that originates from the bifurcation of the Kirchhoff vortex at $r = 3$.

If we start with a small translating pair and allow their sizes to increase quasi-steadily, while keeping the linear momentum (i.e. the separation ℓ) constant, the solution may or may not fail to exist for finite A. But if they are rotating, there is definitely a maximum size for them to exist as a separate pair and we would expect coalescence or fusion to occur at some stage.

The stability of the configurations to two-dimensional disturbances can be discussed qualitatively [20] by applying Kelvin's criterion. It is plausible that the energy of a translating pair is a maximum if only symmetrical isovorticial disturbances are allowed but is a minimax with respect to asymmetrical ones. It is therefore expected that the vortex pair of uniformly translating vortices is in general unstable and the motion becomes unsteady, unless the disturbances are constrained to be symmetrical by, for example, introducing a splitter plate so that each vortex interacts with its image, in which case the configuration is stable (at least to two-dimensional disturbances). Barker (unpublished) finds experimentally that this is indeed the case. For the rotating pair, the excess energy is an absolute minimum for $A/\ell^2 < 5.00$ on the branch coming from infinity, so the pair is stable until the limit point is reached and the continuation past the limit point is unstable. This result agrees well with numerical solutions of the initial value problem for a pair of initially circular uniform vortices.

Analysis of three-dimensional stability is limited to the atomic representation, using the Biot-Savart law, valid for $A/\ell^2 \ll 1$ and $A/\lambda^2 \ll 1$ where λ is the wavelength. Crow [4] found the translating pair is weakly unstable while Jimenez [6] found the rotating pair is stable, under these restrictions. The weak short wave parametric instability is presumably also present when $A/\ell^2 \ll 1$.

6. The Linear Array of Uniform Vortices.

We now describe the results for the equilibrium shapes of an infinite, straight array of equal uniform vortices of area A and separation L [21, 17]. This flow is a molecular representation of the turbulent mixing layer. The shapes of the vortices depend on only one parameter $A^{\frac{1}{2}}/L$; the strength Γ of the vortices just scales the velocity field. At infinity, the velocity is $\pm U = \pm \frac{1}{2}\Gamma/L$.

For small $A^{\frac{1}{2}}/L$, the solution is unique and the vortices are approximately circular. As $A^{\frac{1}{2}}/L$ increases, the vortices become more deformed, taking elliptical shapes with major axis parallel to the line of centers. Following the solutions as a function of $A^{\frac{1}{2}}/L$, one finds a limit point, i.e. vortex of maximum area at $A^{\frac{1}{2}}/L = 0.4875$. Past the limit point, the length a of each vortex continues to increase, although the area is decreasing, until the vortices touch when $A^{\frac{1}{2}}/L = 0.478$. The maximum value of the width b

is reached before the limit point. The solutions can be continued into a family of connected vortices or varicose vortex sheets which ends in a uniform vortex sheet of width 0.2L.

An average velocity profile $U(y)$ can be constructed, defined by

$$U(y) = \frac{1}{L} \int_0^L u(x,y)\,dy \,, \tag{6.1}$$

from which a vorticity width $\delta_\omega = 2U/U'(0)$ can be defined. For $A^{\frac{1}{2}}/L > 0.3$, δ_ω/L is in the range 0.25 - 0.3, which incidentally is in the range of the experimental observations.

The stability to two-dimensional disturbances can be investigated qualitatively using Kelvin's criterion. For this purpose, we need the excess energy E per unit length, shown in figure 1, normalised on $L = 1$, $\Gamma = 1$, as a function of $\theta = A^{\frac{1}{2}}/L$. The dimensional excess energy is

$$\mathcal{E} = 4\,U^2\,L\,E(\theta) \,. \tag{6.2}$$

It is necessary to distinguish between superharmonic and subharmonic disturbances. The former keep the period at L, the latter have period nL, $n > 1$. It can be argued that the excess energy is bounded above, but can be made arbitrarily large and negative, for isovorticial perturbations of period L. Hence the upper branch in figure 1, $0 < A^{\frac{1}{2}}/L < 0.4875$, is stable to superharmonic disturbances, whereasethe lower branch is a minimax of excess energy and unstable. Thus the limit point is an exchange of stability.

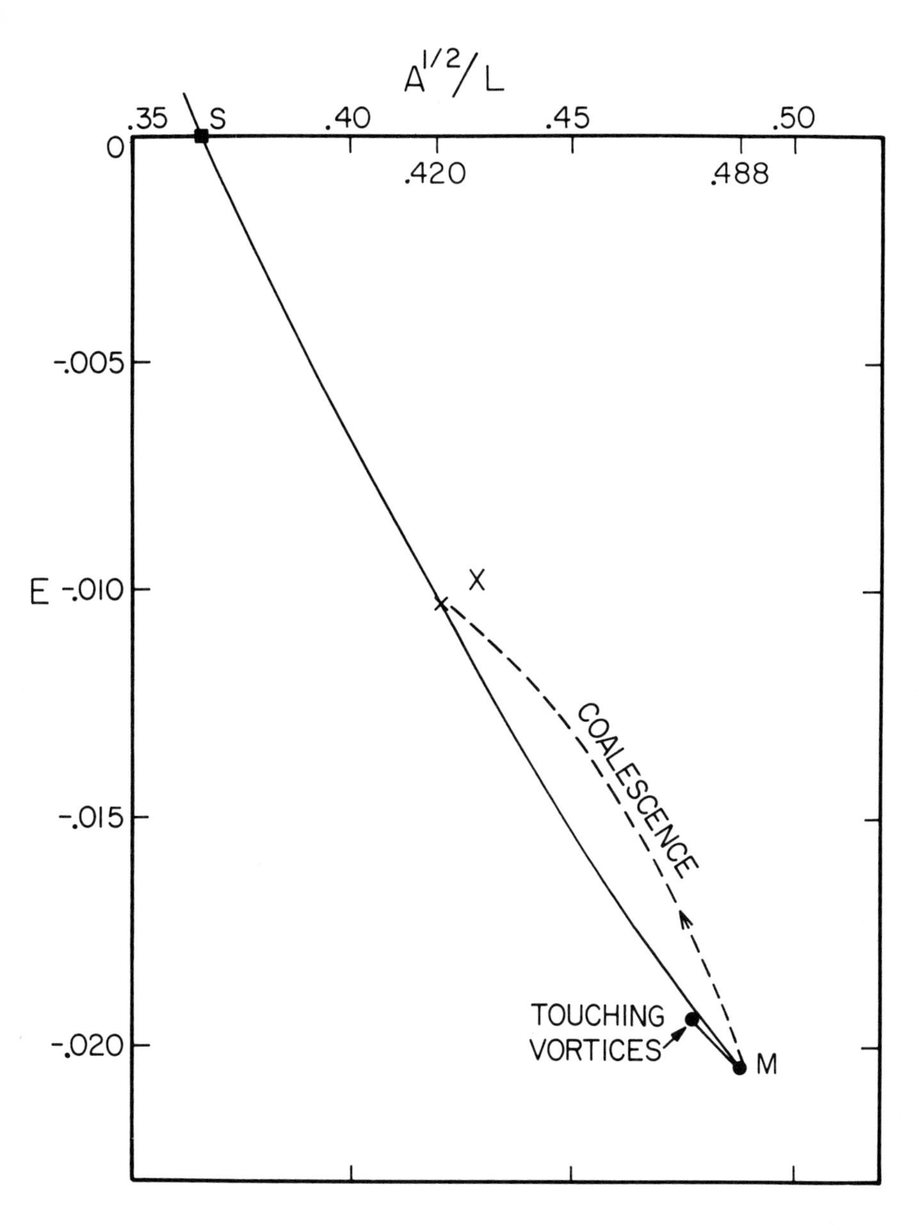

Figure 1. Dimensionless excess energy E as a function of vortex size. Energy is measured relative to a vortex sheet of same average strength. X denotes state obtained by coalescence (pairing) of state M of vortices of maximum area without loss of energy but with ingestion of irrotational fluid into the vortices.

But subharmonic disturbances are always unstable, for if $L \to nL$ $A \to nA$, and $\theta \to \theta/n^{\frac{1}{2}}$, the energy is increased since it is easily seen that

$$n\,E\left(\frac{\theta}{n^{\frac{1}{2}}}\right) > E(\theta) \qquad (6.3)$$

for all θ. Thus we expect that the array is always unstable to the pairing instability, which can be calculated exactly (unpublished) in an atomic representation $A = 0$. If this particular representation of the turbulent mixing layer has merit, it implies that the evolution of the layer is a race between the destabilisation to superharmonic disturbances at finite size, the so-called 'tearing' process, and the destabilisation to subharmonic disturbances, the 'pairing' process. We comment further on this in §7.

The flow and stability can be calculated exactly for hollow vortices [21, 2] and the conclusions from Kelvin's criterion verified explicitly.

Very little can be said at present about the stability to three-dimensional disturbances. For $A^{\frac{1}{2}}/L << 1$, an atomic representation could be used to study stability to long wavelength disturbances. This would be worth doing but has apparently not yet been carried out. The weak parametric instability should also be present, and there is evidence of something like it in the experiments, but there seem to be some differences in detail and it is too early to say that it explains the observed spanwise instability.

It should be mentioned that Pierrehumbert [16] has studied the stability to two- and three-dimensional disturbances of the steady exact solution of the Euler equations described by

$$\psi = \frac{1}{2}\log(\cosh 2y - \rho\cos 2x) \qquad (6.4)$$

with $0 < \rho < 1$. As ρ changes from 0 to 1, this flow field goes from a hyperbolic tangent profile to an array of point vortices. A good overall fit with experimental profiles is claimed for $\rho = 0.25$. The results seem consistent with those derived by qualitative arguments for the array.

7. Coherent Structures.

The question to be considered now is the relevance of these exact results to the coherent structure concept in turbulence. Of course, there are more molecular representations that need to be studied before a complete answer can be given. For example, the Karman vortex street is an obvious subject for application to two-dimensional turbulent wakes (and is currently under study). An array of vortices lying at the vortices of a regular polygon would provide information about the effect of curvature on a mixing layer. Arrays of ring vortices should be investigated for application to axisymmetric jets and wakes, or helical vortices, and so on. The interaction of ring vortices is a fascinating problem, but the theoretical difficulties are great and little progress has been made. Norbury [14] has computed the shapes of steady vortex rings of arbitrary core size.

The answer we can give so far is that the analysis gives qualitative support to the concept of quasi-permanent coherent structures. The fact that such solutions exist and are not strongly unstable implies that organized vortices can have finite lifetimes, and they also provide a useful framework in which to describe experimental results.

As an example of the type of prediction that can be made, we discuss the mixing layer and the results of §6. The mixing layer is represented at some station by one of the steady exact solutions of the Euler equations. Then the actual effects of turbulence can be modelled by an increase in the area of the vortices, the separation between them staying constant. The state of the mixing layer can then be represented by points on the E vs $A^{\frac{1}{2}}/L$ curve of figure 1. There are two possibilities, which can be categorized as pairing or tearing. In the pairing mode, we assume that the energy of the layer is conserved during the entire evolution, in which case it is necessary that $E = 0$ and the instantaneous state is described by the point S on figure 1. Then

$$\frac{a}{b} = 1.45, \quad \frac{\delta_\omega}{L} = 0.28 \quad . \tag{7.1}$$

In this case the vortices would not grow between pairings, i.e. there is no turbulent entrainment or 'nibbling', and in the pairing process the vortices would have to ingest or 'gulp' a volume of fluid equal to themselves in order to maintain the similarity with $A \propto L^2$. The time scale of the evolution would be the pairing time, which could be estimated from the exact solution for point vortices if there were a reliable way of estimating the stage at which coalescence takes place. The results of §5 could perhaps be applied to this question. According to the exact solution for the finite amplitude pairing instability of a row of point vortices, the time to closest approach is $(4L^2/\pi\Gamma)\ln(2/\pi\varepsilon)$, where εL is the initial displacement.

The second possibility is the tearing process, according to which the system moves from some state X by nibbling until the maximum area is attained at M. The steady state can no longer exist and something catastrophic occurs; presumably the vortices disintegrate and reform into a new array with double the spacing. The point X can be estimated by the assumption that energy is conserved during the rearrangement. Then θ_X satisfies

$$2\,E(\theta_X) = E(\theta_M) \tag{7.2}$$

and hence $\theta_X = 0.420$. There is then a 35% increase in volume during the movement from X to M and the remaining 65% is gulped during the reformation. The time scale of the mixing layer evolution is then fixed by the nibbling rate. The dimensions are

$$\left.\begin{aligned} \left(\frac{a}{b}\right)_X &= 1.54\,, \qquad \left(\frac{a}{b}\right)_M = 2.4 \\ \left(\frac{\delta_\omega}{L}\right)_X &= .30, \qquad \left(\frac{\delta_\omega}{L}\right)_M = .27\ . \end{aligned}\right\} \tag{7.3}$$

At present, the experimental evidence does not clearly distinguish between these two processes. It is possible that pairing is dominant when the layer is young, and tearing becomes more important as the layer ages, because once a significant amount of energy is lost, the pairing process would have to dissipate energy suddenly in order to preserve the similarity as the negative excess energy is doubled. Of course, the analysis does not explain why the vortices form in the first place or survive the numerous interactions rather than forming an amorphous cloud. Perhaps Onsager's negative temperature concepts are needed to explain this.

REFERENCES

1. Arnold, V.I., Mathematical Methods of Classical Mechanics, Springer-Verlag, New York, 1980.
2. Baker, G.R., Saffman, P.G. and Sheffield, J.S., Structure of a linear array of hollow vortices of finite cross section, J. Fluid Mech. 74 (1976) 469-476.
3. Christiansen, J.P. and Zabusky, N.J., Instability, coalescence and fission of finite-area vortex structures, J. Fluid Mech. 61 (1973) 219-243.
4. Crow, S.C., Stability theory for a pair of trailing vortices, A.I.A.A. J. 8 (1970) 2172-2179.
5. Deem, G.S. and Zabusky, N.J., Vortex Waves; stationary V states, interactions, recurrence and breaking, Phys. Rev. Lett. 40 (1978) 859-862.
6. Jimenez, J., Stability of a pair of co-rotating vortices, Phys. Fluids 18 (1975), 1580-1581.
7. Lamb, H., Hydrodynamics, Cambridge University Press 1932.
8. Leonard, A., Vortex methods for flow simulation, J. Comp. Phys. 1980 (to appear).
9. Love, A.E.H., On the stability of certain vortex motions, Proc. Lond. Math. Soc. (1) 25 (1893), 18-30.
10. Milinazzo, F. and Saffman, P.G., The calculation of large Reynolds number two-dimensional flow using discrete vortices with random walk, J. Comput. Phys. 23 (1977), 380-392.
11. Moore, D.W., The velocity of a vortex ring with a thin core of elliptical cross section, Proc. Roy. Soc. A 370 (1980) 407-415.
12. Moore, D.W. and Saffman, P.G., Structure of a line vortex in an imposed strain, in Aircraft Wake Turbulence (Eds. Olsen, J.H., Goldburg, A. and Rogers, M.) Plenum Press, New York, 1971, 339-354.

13. Moore, D.W. and Saffman, P.G., The instability of a straight vortex filament in a strain field, Proc. Roy. Soc. A 346 (1975), 413-425.
14. Norbury, J., A family of steady vortex rings, J. Fluid Mech. 57 (1973), 417-431.
15. Pierrehumbert, R.T., A family of steady translating vortex pairs with distributed vorticity, J. Fluid Mech. 99, 129-144.
16. Pierrehumbert, R.T., The structure and stability of large vortices in an inviscid flow, M.I.T. Fluid Dynamics Lab. Rep. 80-1 (1980).
17. Pierrehumbert, R.T. and Widnall, S.E., The structure of organized vortices in a shear layer, A.I.A.A. Paper 79-1560 (1979).
18. Saffman, P.G., Coherent structures in turbulent flow, Conference on the role of coherent structures in modelling turbulence and mixing, Madrid 1980. Lecture Notes in Physics. Springer-Verlag (to appear).
19. Saffman, P.G. and Baker, G.R., Vortex interactions, Ann. Rev. Fluid Mech. 11 (1979), 95-122.
20. Saffman, P.G. and Szeto, R., Equilibrium shapes of a pair of equal uniform vortices, Physics of Fluids (to appear 1981).
21. Saffman, P.G. and Szeto, R., Structure of a linear array of uniform vortices, Stud. in Appl. Math (to appear 1981).
22. Zabusky, N.J., Hughes, M.H. and Roberts, K.V., Contour dynamics for the Euler equations in two dimensions, J. Comput. Phys. 30 (1979), 96-106.

The author was supported by the Army Research Office Durham, (DAAG 28-78-C-0011) and the Department of Energy (Office of Basic Energy Sciences). The author's computations on vortex arrays and vortex pairs were carried out on the CDC Cyber 203 Computer at the CDC Service Center, Arden Hills, Minnesota and the granting of the computing time is gratefully acknowledged.

Applied Mathematics 217-50
California Institute of Technology
Pasadena, California 91125

Interactions between Large-Scale Coherent Structures and Fine-Grained Turbulence in Free Shear Flows

J. T. C. Liu

1. INTRODUCTION

I should like to begin by quoting some remarks made by Hans Liepmann in his 1952 survey paper on aspects of the turbulence problem [30] in connection with turbulent flows with secondary structure:

> In recent years the importance of the existence of a secondary, large-scale structure in turbulent shear flow has become apparent. Corrsin [11]* and Townsend [60] found that the flow near the outer edge of a jet or wake is only intermittently turbulent. ...
>
> While the large eddies ordinarily found in intermittently turbulent flow appear statistically distributed, there do exist a number of cases in which a regular or nearly regular motion of large-scale superimposed upon turbulent flow has been observed. Pai [44] and MacPhail [38] found that the three-dimensional vortices which initiate the laminar instability in the flow between rotating cylinders (Taylor [57]) persist even if the flow has become fully turbulent. In recent measurements on a vortex street behind a cylinder Roshko [49] found a similar result: above a cylinder Reynolds number of about 150 the wake flow is essentially turbulent with superimposed, nearly equally spaced vortices. ...
>
> ... While there may be doubt about the details of this secondary structure, there is no doubt that the large-scale motion cannot be ignored for a great many problems including problems in sound production from jets, combustion, etc.

*The reference numbers are those in the present paper.

ISBN 0-12-493240-1

These rather perceptive observations serve well as introductory remarks to a paper on the large-scale structure in free turbulent flows in 1980!

The more recent observational evidence of large-scale coherent structures in free shear flows is strikingly displayed in the work of Brown and Roshko [8] for the mixing region. In a work which is similar to that of Roshko [49], Taneda [56] showed that the Kármán vortex street behind a cylinder broke up into fine-grained turbulence but that the vortex street reformed further downstream in a turbulent flow; this subsequently also broke down and reformed but with the scales becoming larger as the flow progressed downstream. Tenada [56] also showed that the wake behind a flat plate has similar features. It appears that a disturbance regeneration process, whether due to the remnants of the original disturbance or due entirely to new disturbances, is inherent in such flows. Turbulent vortex streets in wakes have also been observed more recently by Papailiou and Lykoudis [45].

In a real, developing free shear flow, the large-scale structures are propagated downstream into regions of variable mean motion. The observed growth and decay of the large-scale structure necessarily imply that the energy production and dissipation change as the structure proceeds downstream, and that they are not necessarily in local balance. The rate at which the large-scale structure grows and decays depends crucially on the imbalances between the rate-controlling factors of production and "dissipation".

In the theoretical discussion that follows, I shall proceed with a "prejudice" that the large-scale structures in turbulent shear flows, particularly free turbulent flows, are a manifestation of hydrodynamic instability. In this situation, the inflexional mean flow, which is dynamically unstable, would provide the production mechanism for the large-scale structure whether the flow is laminar or "turbulent". The similarity in the production mechanism accounts for the presence of large-scale coherent structures in the very high Reynolds-number situation where fine-grained turbulence is present as well as in the lower Reynolds-number range where the basic flow is still laminar. However, this does not imply that the fine-grained turbulence has no effect on the large-scale structure.

Although much can be learned about the properties of large-scale structure in turbulent flows from studying such structures in laminar flows, one of the most important rate-controlling mechanisms is missing from such studies. The straining of the fine-grained oscillations by the passage of the large-scale structure would provide the mechanism for energy exchange between the two disparate scales of motion; this energy exchange, as intuition might tell us, is dissipative for the large-scale structures. The extent of this dissipative role, of course, depends on the energy levels of the fine-grained turbulence. The interpretation of the large-scale structures as being the result of hydrodynamic instability would also imply that the ability of these structures to extract energy from the mean flow is spectrally dependent--some modes are more efficient than others. In addition, the straining of the fine-grained turbulence and the resulting energy transfer between the disparate scales of motion would be spectrally dependent as well. In developing free flows, the occurrence of large-scale structures is, in general, a nonequilibrium event in that decay and growth takes place.

I should like to return to the existence of Taylor vortices and the like in the turbulent flow between rotating cylinders. The instability mechanism there is the competition of centrifugal forces and of dissipative mechanisms--be they molecular viscosity (as in the laminar case) or fine-grained turbulence. This instability mechanism is quite different from that of the dynamical instability in free shear flows and from the Tolmein-Schlichting-type instability of wall-bounded flows where "viscous" effects rearrange the phase shift between the disturbance velocity components and give rise to a Reynolds shear stress that is favorable to the extraction of energy from the mean flow. It is conceivable that one would neither use second-order Reynolds stress "closure" methods to "predict" such instabilities in the laminar problem, nor expect that a single set of nonvariable closure constants could describe instabilities due to such different physical mechanisms. From this point of view, then, the lack of success of second-order closure models in turbulent shear flows may be attributable to the non-universal behavior of large-scale coherent structures primarily because of the non-universality of their instability mechanisms.

I appreciate the difficulties in observing experimentally features of the large-scale structure in turbulent flows, the situation being aggravated by the "jittering" of the phases of such structures (see, for instance, Thomas and Brown [58]). This growth and decay process* as well as the simultaneous occurence of several such events of other spectral components would become obscured in the correlation measurements. Some time ago, Liepmann [31] suggested that the large-scale structures in turbulent shear flows ought to be studied in a well-controlled manner, in much the same spirit as the Schubauer and Skramsted [53] experiments, and that the large-scale structure could be considered as a "secondary" structure superimposed upon a mean flow and the background fine-grained "turbular" fluid (as discussed by Townsend [61]). However, it was not then clear how the large-scale structure fluctuations could be sorted out, either experimentally or theoretically, from the total fluctuations.

The imposition of a fixed-phase disturbance on a turbulent shear flow and the use of the phase average, essentially a "conditional average", enables one to sort out the large-scale structure from the random fine-grained turbulence. This augments the usual Reynolds [47] averaging procedure which sorts out the mean flow from the total fluctuations. The experimental realization of such methods was discussed by Hussain and Reynolds [21], Kendall [25], Binder and Favre-Marinet [5] and Favre-Marinet and Binder [15]. In order to fix ideas with respect to the physical picture which the subsequent theoretical considerations hope to explain, we refer to the experimental results depicted by Figure 10a of the last reference, reproduced here as Figure 1. Favre-Marinet

*In confined, nondeveloping flows, such as that between rotating cylinders, it appears that "stationary" (as contrasted with "instantaneous") equilibrium can exist between the coherent structures and turbulence, although the equilibrium states themselves appear to take on multiple values.

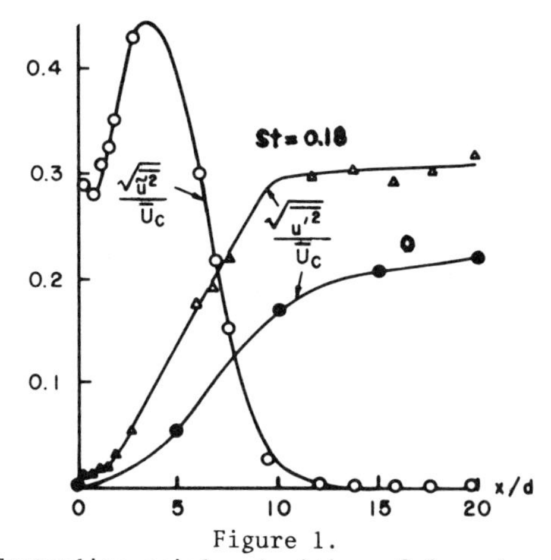

Figure 1.
Centerline axial velocities of forced and unforced jet (Favre-Marinet and Binder [15]).

and Binder imposed a well-controlled large-scale structure on a round jet and took hot wire measurements on the jet axis. In the subsequent development, the large-scale structure amplified and then decayed, whereas the fine-grained turbulence level increased at a faster rate than it did when an imposed large-scale structure was absent. In the experiments of Figure 1, the signals were normalized by the local mean centerline velocity. The speading of the jet was enhanced by the forcing, with an accompanying more rapid decay of the mean centerline velocity. In this case, it appears that the "burst" of fine-grained turbulence follows the demise and occurs at the expense of the large-scale structure. The previous growth of the large-scale structure occurs by energy extraction from the mean flow. At this stage, then, it is not inconceivable that the large-scale structure plays an intermediary role in negotiating an indirect transfer of energy from the mean flow to the fine-grained turbulence.

In the following, I shall discuss some of the "practical" ways of modeling the large-scale structure and its interaction with fine-grained turbulence; the primary purpose of such models is, of course, to try to understand the observed growth and decay processes in the nonequilibrium development of the large-scale structure. This will necessarily involve the synthesis of numerous ideas due to many of the participants as well as nonparticipants in this symposium, including, of course, those of the chairman of this session [39].

2. SOME REMARKS ON AVERAGING AND THE BASIC EQUATIONS

Following Townsend [61], we split the total flow quantity $q(x_i t)$, where x_i are the spatial coordinates and t is the time, into the three components

$$q = \overline{Q} + \tilde{q} + q',$$

where $\overline{Q}$ denotes the Reynolds mean flow, $\tilde{q}$ the large-scale structure and q' the fine-grained turbulence. In the Reynolds framework $(\tilde{q}+q')$ would be considered as turbulence, here $\tilde{q}$ is considered as periodic with a fixed phase. In the laboratory, as in the mixing layer experiments of Brown and Roshko [8], this periodicity is in time with a fixed period T and the development or evolution is in the downstream direction (denoted by the horizontal variable x). The shear layer spreads in the vertical direction (denoted by z) and is essentially two-dimensional and independent of the span (y is used to denote the spanwise variable). In the tilting tube experiment (Thorpe [59]), where a long tube is filled with a heavier fluid on the bottom and a lighter fluid on the top, a slight tilt sets up a shear layer that is homogeneous in the "horizontal" direction and spreads vertically uniformly in the mean. Instabilities in this situation are periodic horizontally and develop in time. These two contrasting situations are known respectively as the spatial and temporal problems in hydrodynamic stability. The linearized problem in the special situation of small amplification rates is the only case in which the temporal and spatial instabilities can be related via the group velocity (Gaster [17]). In general, for nonlinear problems, there is no valid transformation which can relate the details of the two problems, although there is a strong qualitative resemblance between them. Computationally the temporal problem is considerably simpler; but for a well-controlled large scale structure the spatial problem appears to be the easier to reproduce in the laboratory.

For the spatial problem with time periodicity T, the Reynolds average (denoted by an overbar) is defined as

$$\overline{q} = \overline{Q} = \frac{1}{T}\int_0^T q\,dt.$$

In this case, the conditional average tied in with a fixed phase (denoted by < >) is [21], [15]

$$\langle q\rangle = \overline{Q} + \tilde{q} = \frac{1}{N}\sum_{n=0}^{N} q(x_i,\ t+nT).$$

For the temporal problem with horizontal spatial periodicity λ, the appropriate Reynolds average is

$$\overline{q} = \frac{1}{\lambda}\int_0^\lambda q\,dx,$$

and the appropriate conditional average in this case is

$$\langle q\rangle = \frac{1}{N}\sum_{n=0}^{N} q(x+n\lambda, y, z, t).$$

Because q' and $\tilde{q}$ are not correlated, the conditional average essentially distinguishes $\tilde{q}$ from the total fluctuations.

Some time ago Amsden and Harlow [2] numerically computed the nonlinear time-development of Kelvin-Helmholtz instability with horizontal periodicity directly from the unsteady Navier-Stokes equations (rather than splitting the dynamical variable into a Reynolds mean and fluctuation with zero mean). The dependent variable thus comprises the total flow quantity and the Reynolds mean can be calculated after the numerical results are obtained in order to study the interactions between the mean motion and the fluctuations.

Patnaik, Sherman and Corcos [46] studied the same laminar problem with stratification. There, in the absence of fine-grained turbulence ($q' = 0$), the dynamical variable is $(\overline{Q}+\tilde{q})$ and is to be obtained from the Navier-Stokes equations without averaging. To apply such ideas to the present problem in the presence of fine-grained turbulence, the conditional average is used to sort out the total coherent signal $\langle q\rangle = (\overline{Q}+\tilde{q})$ from the fine-grained turbulence. The modulated Reynolds stresses, $\langle q'q'\rangle = (\overline{q'q'} + \tilde{r}_{q'q'})$, which contains a Reynolds mean contribution and a contribution induced by the large-scale structure, augments the viscous diffusion in the present problem.

To be more specific, the Navier-Stokes equations for an incompressible, homogeneous fluid are split into a coherent part and a random fine-grained turbulence part with the velocity components and pressure written as

$$u_i = U_i + u_i', \quad p = P + p',$$

where U_i, P consist of the Reynolds mean $\overline{U}_i, \overline{P}$ and the coherent fluctuation part $\tilde{u}_i, \tilde{p}$ which has zero Reynolds mean. The modulated stresses then take the form $\langle u_i'u_j'\rangle = \overline{u_i'u_j'} + \tilde{r}_{ij}$. We have seen that the coupled conservation equations for U_i, P and the transport equations for $\langle u_i'u_j'\rangle$ are identical in form to those for the Reynolds [47] system for $\overline{U}_i, \overline{P}$ and $\overline{(u_i'+u_j')(u_j'+u_j')}$ [19, 34]

$$\frac{\partial U_i}{\partial x_i} = 0, \tag{2.1}$$

$$\frac{\partial U_i}{\partial t} + \frac{\partial U_i U_j}{\partial x_j} = -\frac{\partial P}{\partial x_i} - \frac{\partial \langle u_i'u_j'\rangle}{\partial x_j} - \frac{1}{Re}\frac{\partial^2 U_i}{\partial x_j \partial x_j}, \tag{2.2}$$

$$\begin{aligned}\left(\frac{\partial}{\partial t} + U_k \frac{\partial}{\partial x_k}\right)\langle u_j'u_k'\rangle = &-\underbrace{\left[\langle u_j'u_k'\rangle \frac{\partial U_i}{\partial x_k} + \langle u_i'u_k'\rangle \frac{\partial U_j}{\partial x_k}\right]}_{\text{production}} \\ &+ \underbrace{\left\langle p'\left(\frac{\partial u_i}{\partial x_j} + \frac{\partial u_j'}{\partial x_i}\right)\right\rangle}_{\text{redistribution}} \\ &- \underbrace{\frac{\partial}{\partial x_k}\left[\langle u_i'u_j'u_k'\rangle + \langle p'(u_i'\delta_{jk} + u_j'\delta_{ik})\rangle - \frac{1}{Re}\frac{\partial \langle u_i'u_j'\rangle}{\partial x_k}\right]}_{\text{diffusion}} - \underbrace{2\frac{1}{Re}\left\langle \frac{\partial u_j'}{\partial x_k}\frac{\partial u_j'}{\partial x_k}\right\rangle}_{\text{dissipation}}\end{aligned} \tag{2.3}$$

where Re is the Reynolds number and all the variables have been made dimensionless by the appropriate characteristic velocity, length scale and density of the problem. The system involves no explicit Reynolds averages, but these could be obtained once U_i and $\langle u_i'u_j'\rangle$ have been computed. However, the closure necessitated by (2.3) could conceivably follow the functional form of second-order Reynolds stress closure schemes because of its similarity to the original Reynolds stress equations. The stresses $\langle u_i'u_j'\rangle$ involves only fine-grained turbulence and are thus much more likely to be universal than $\overline{(\tilde{u}_i+u_i')(\tilde{u}_j+u_j')}$.

Of course, conservation equations can always be obtained for $\overline{U}_i$ and $\tilde{u}_i$ and for $\overline{u_i'u_j'}$ and $\tilde{r}_{ij}$ [21], but such equations are too cumbersome to state here. The interactions between the mean flow, large-scale structure and fine-grained turbulence will be illustrated through the Reynolds averaging of the results obtained from the numerical computation of a simplified problem in Section 3. Subsequently, we shall discuss the interaction problem through the use of approximated forms of the conservation equations for the three components, following some earlier ideas [32, 34] in modeling the nonlinear interaction processes.

3. NUMERICAL COMPUTATIONS APPLIED TO A SIMPLE PROBLEM

In this section we shall discuss the application [19] of the framework discussed in Section 2 to the Kelvin-Helmholtz wave developing in a shear flow with fine-grained turbulence, a problem considered by Amsden and Harlow [2] for an originally laminar shear layer. Here it is considered that the Reynolds-mean shear layer, which consists of oppositely directed streams, is horizontally homogeneous and has coexisted with the fine-grained turbulence for a sufficiently long time that all mean quantities are developing in a self-similar fashion and the shear layer is spreading at a constant rate. A large-scale structure is imposed at $t = t_0$, one which is hydrodynamically consistent with the prevailing mean velocity. The large-scale structure is considered as two-dimensional (or dominantly two-dimensional) with vorticity axis in the spanwise direction. Both the spanwise conditionally averaged velocity and spanwise derivatives of conditionally averaged quantities vanish.

The conditionally averaged stream function is defined by

$$U = \frac{\partial \Psi}{\partial z}, \quad W = -\frac{\partial \Psi}{\partial x}, \tag{3.1}$$

where U,W and x,z are the dimensionless horizontal and vertical velocities and coordinates respectively. The characteristic velocity here would be that of the free stream and the length scale that of the initial $(t = t_0)$ thickness of the free shear layer. The stream function here is also related to the conditionally averaged vorticity

$$\Omega = -\nabla^2\Psi, \tag{3.2}$$

where ∇^2 is the Laplacian in the x,z-plane. The nonlinear vorticity equation, written in terms of the stream function, is

$$\nabla^2\Psi_t + \Psi_z\nabla^2\Psi_x - \Psi_x\nabla^2\Psi_z = \langle u'w'\rangle_{xx} - \langle u'w'\rangle_{zz} + (\langle w'^2\rangle - \langle u'^2\rangle)_{xz}, \tag{3.3}$$

where partial differentiation is indicated by the appropriate subscripts. The vertical boundary conditions require that far away from the shear layer all "disturbance" quantities vanish and that $\Psi_z \to \bar{U} = 1$. Horizontal periodic boundary conditions are applied to all conditionally averaged quantities, the domain being a multiple of the large-scale structure wavelength. We shall limit the discussion in this section to a domain of one wavelength, restricting the interactions to occur between the large-scale structure and the fine-grained turbulence. We shall discuss, at a later stage, the two-wavelength domain and Kelly's mechanism [24].

We have neglected the viscous diffusion $\nabla^4\Psi/Re$ in (3.3) in comparison to the more vigorous effects of the transport by the turbulence stresses. We might remark here that retention of viscous diffusion in (3.3) could form a framework for the study of the incipient transition from laminar to "turbulent" flow where the fine-grained turbulence, (such as would exist naturally in a wind tunnel) appears initially "weak". The straining of such weak turbulence by the coherent instability structures eventually allows the fine-grained turbulence to grow at the expense of the coherent instabilities. From this point of view, the transition from laminar to "turbulent" flow and the evolution of coherent large-scale structures in a turbulent flow can be somewhat unified, depending on the original levels of the fine-grained turbulence. The nonlinear vorticity equation (3.3) is coupled to the transport equations for the turbulent stresses and is as yet unaddressed in the closure problem. We have alluded earlier to the fact that the conditional-averaged system appears identical in form to Reynolds [47] original system. The functional forms for the closure scheme here could conceivably follow those for the second-order closure of the Reynolds stresses in the Reynolds [47] system. This was discussed in [19]; it suffices to mention here that the closure schemes of Launder, Reece and Rodi [29] were used. This involves using the approximation for the pressure-velocity strain redistribution that appropriately account for the "rapid distortion" due to the large-scale structure, a gradient diffusion approximation for the triple correlations and a separately modeled transport equation for the rate of viscous dissipation. The net result is that, in addition to Ψ (and Ω), there are five other dependent variables: the shear and normal stresses $\langle u'w' \rangle$, $\langle u'^2 \rangle$, $\langle w'^2 \rangle$ and $\langle v'^2 \rangle$, and the rate of viscous dissipation $\langle \varepsilon \rangle$ accompanied by their their respective transport equations.

We refer to [19] for a thorough discussion of the ini-alization process. Here, the initial conditions for the computations are: the initial turbulence kinetic energy content was $E_{t0} = 1.20 \times 10^{-2}$ and that of the large-scale structure $E_{\ell 0} = 10^{-4}$ and the most amplified mode has wave number $\alpha \approx 0.275$ for the initial mean velocity profile (see Equations (3.5) and (3.6) for definitions of E_ℓ and E_0.

The stream function Ψ computed in [19] is shown in Figure 2 for a series of dimensionless times. The coordinates have now been rescaled by the wavelength of the most aplified mode (corresponding to the initial mean velocity profile here). Except for some details and the rate of development, the patterns of Ψ strongly resemble what would occur in a laminar flow. The vorticity Ω is shown in Figure 3. The initial profile has two cells of opposite sign which agglomerated rapidly, but eventually multiple cells appear to have developed and reagglomerated. Comparison of the vorticity contours with the streamline patterns reveals that there are strong vorticity nonuniformities within the "cat's eye". This is, however, not unexpected as the

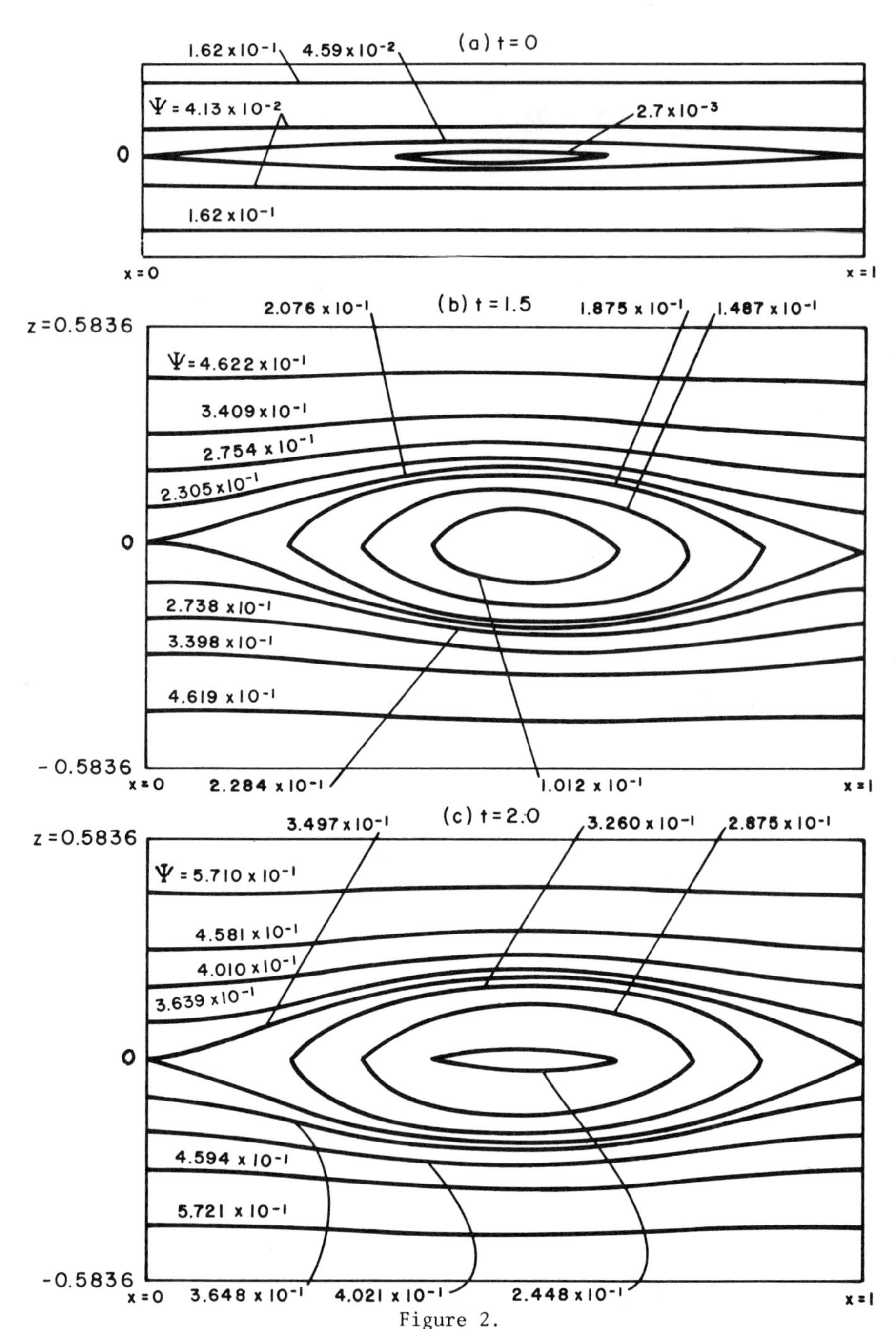

Figure 2.
Evolution of the total coherent stream function with time.

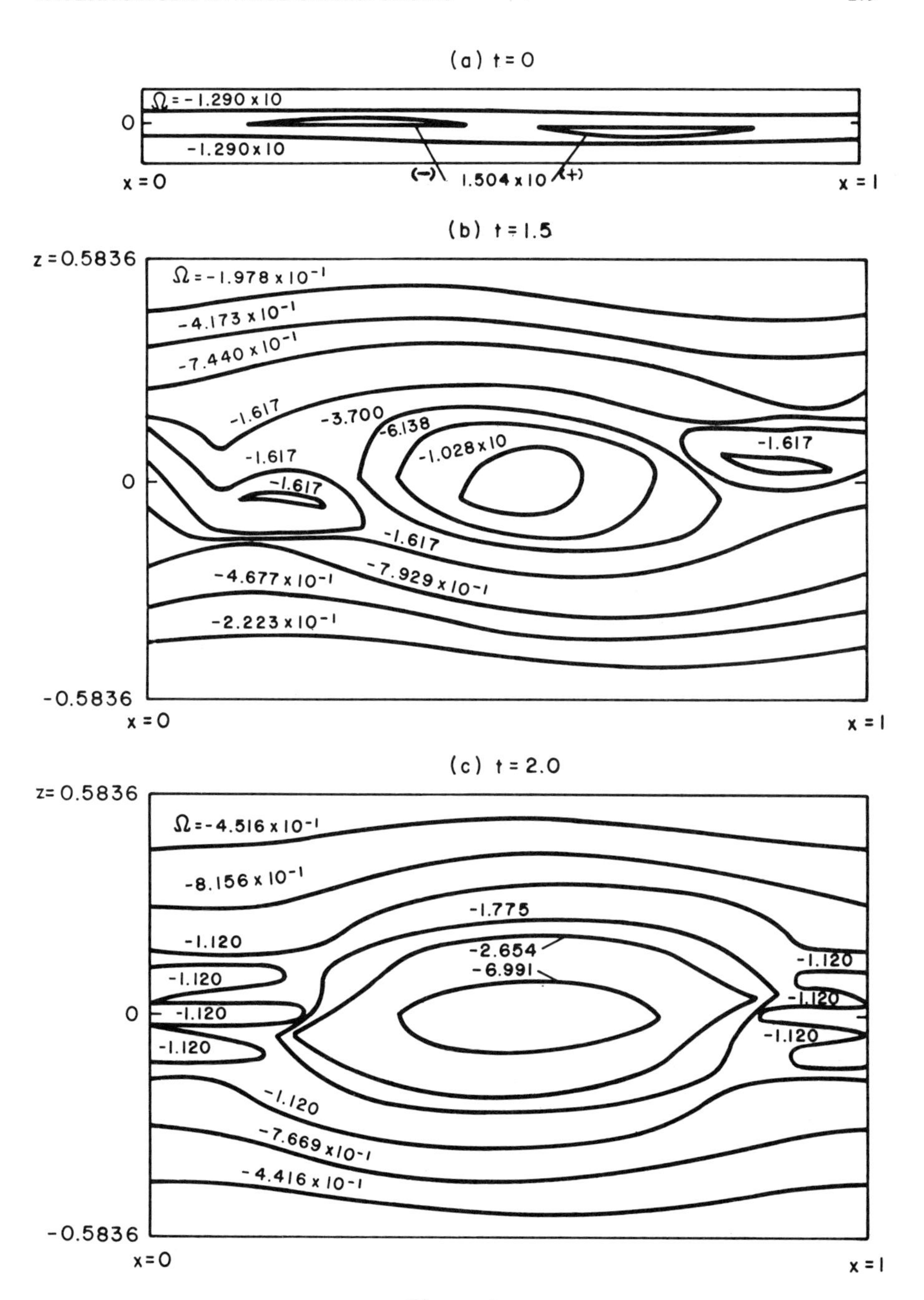

Figure 3.

Evolution of the total coherent vorticity with time.

nonequilibrium development here is a $t = 0(1)$ problem. In a nonlinear critical-layer theory for the present problem, the vorticity would be required to be uniform within the "cat's eye" and this would be achieved as $t \to \infty$ as the fine-grained turbulence smoothed out the vorticity distribution. However, at that stage the fine-grained turbulence will have already achieved a complete dissolution of the large-scale structure. Whether a nonlinear critical-layer theory for the present class of problems can be constructed remains an open question.

The large-scale structure is two-dimensional here, and there is direct conversion to the horizontal and vertical contributions to the turbulence kinetic energy. For the $\langle u'^2 \rangle/2$ component, the mechanisms come from the work done by the normal stress against the horizontal rate of strain $-\langle u'^2 \rangle \partial U/\partial x$ and that by the shear stress against the shearing rate of strain $-\langle u'w' \rangle \partial U/\partial z$. The latter mechanism appears to be the dominant one; Figures 4a and 4b show this dominant source and $\langle u'^2 \rangle/2$. We note that there are pockets of conversion of turbulence energy back to the mean flow, but this appears to be a minor effect. The vertical contribution to the turbulence kinetic energy $\langle w'^2 \rangle/2$ comes from the normal stress production $-\langle w'^2 \rangle \partial W/\partial Z$ and the shearing stress production $-\langle u'w' \rangle \partial W/\partial x$. The shearing stress mechanism again dominates. The patterns of $\langle w'^2 \rangle/2$ are similar to $\langle u'^2 \rangle/2$, and are not shown here. These are the direct turbulence production mechanisms. The spanwise turbulence kinetic energy $\langle v'^2 \rangle/2$ is produced via the isotropizing mechanism of the pressure-velocity strain correlation $\langle p' \partial v'/\partial y \rangle$. In this case, $\langle v'^2 \rangle/2$ is also produced from the two-dimensional large-scale structure through this indirect mechanism. Both $\langle v'^2 \rangle/2$ and $\langle p' \partial v'/\partial y \rangle$ are shown in Figures 5a and 5b, respectively. The present framework can also be extended to allow for the simultaneous presence of streamwise, logitudinal streaks such as those observed by Konrad [28]. In this case, the spanwise large-scale structure provides the additional straining of the fine-grained turbulence giving rise to the direct production of $\langle v'^2 \rangle/2$.

The interactions between the Reynolds-mean flow,

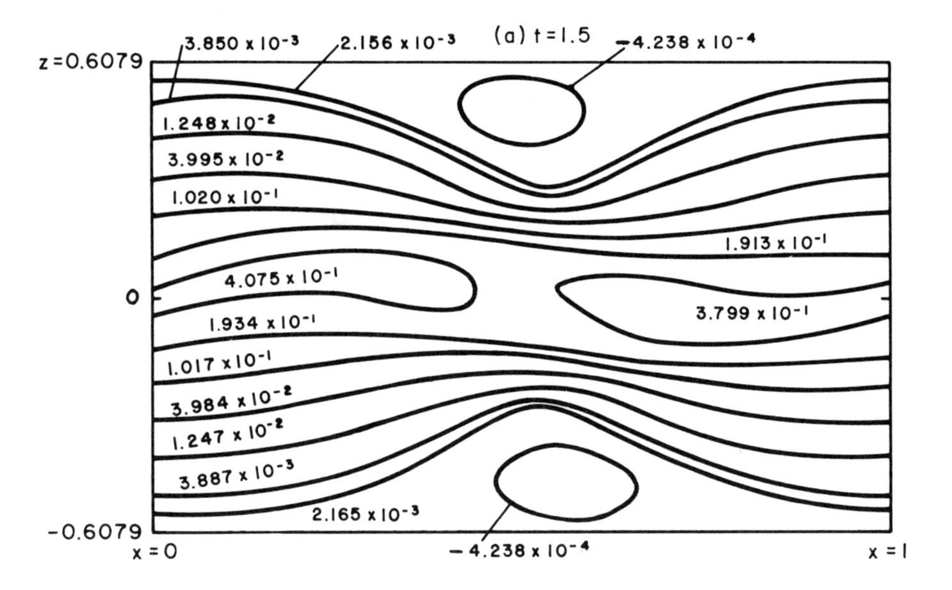

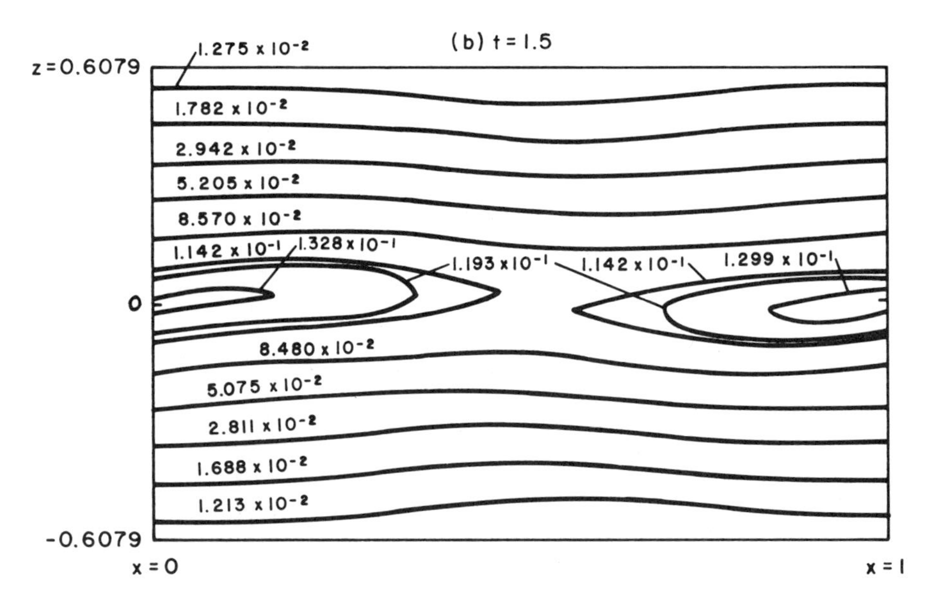

Figure 4.

The horizontal component of condtionally-averaged turbulence kinetic energy and its dominant production mechanism at $t = 1.5$. (a) $-\langle u'w' \rangle \partial U/\partial z$, (b) $\langle u'^2 \rangle /2$.

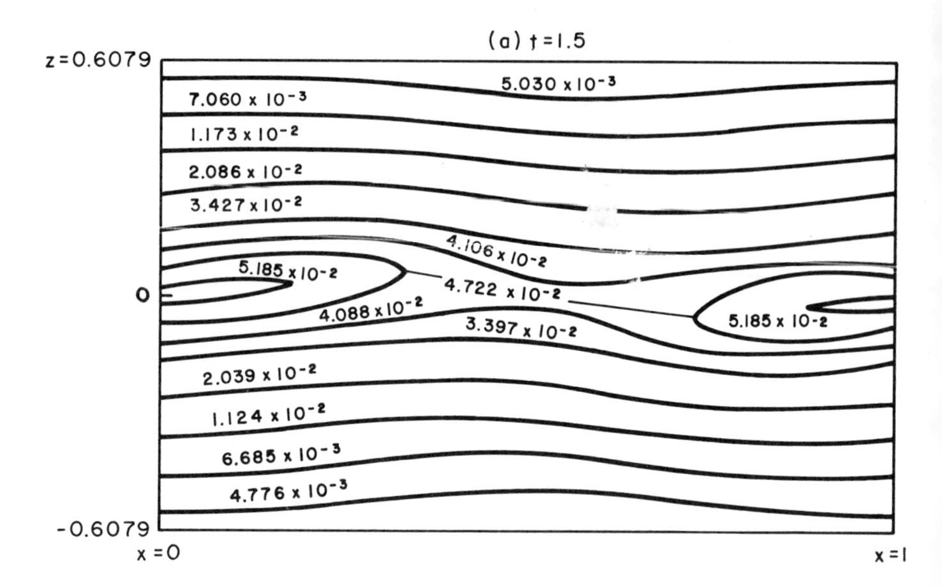

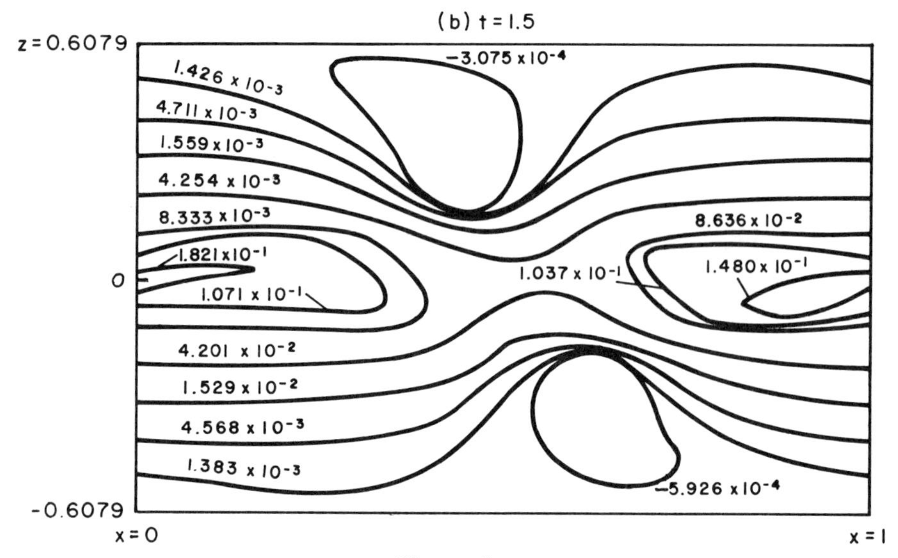

Figure 5.

The spanwise component of conditionally-averaged turbulence kinetic energy and its "production" mechanism at t = 1.5. (a) $\langle v'^2 \rangle/2$, (b) $\langle p' \partial v'/\partial y \rangle$.

large-scale structure and the fine-grained turbulence is then studied by performing the Reynolds average (the horizontal average here) upon the numerical results. The nonequilibrium nature of the interactions is simply and physically illustrated by considering the energy content of the interacting components of flow across the shear layer. The mean flow kinetic energy defect is denoted by

$$E_m = \int_{-\infty}^{0} (\overline{U}^2 - \overline{U}_{-\infty}^2)\,dz + \int_{0}^{\infty} (\overline{U}^2 - \overline{U}_{\infty}^2)\,dz, \quad (3.4)$$

where $\overline{U}$ is the mean flow velocity and $\overline{U}_{\pm\infty} = \pm 1$ are the dimensionless velocities of the upper and lower streams, respectively. The kinetic energy content of the two-dimensional large-scale structure is

$$E_\ell = \frac{1}{2}\int_{-\infty}^{\infty} \overline{(\tilde{u}^2 + \tilde{w}^2)}\,dz, \quad (3.5)$$

and that of the fine-grained turbulence is

$$E_t = \frac{1}{2}\int_{-\infty}^{\infty} \overline{(u'^2 + v'^2 + w'^2)}\,dz. \quad (3.6)$$

The kinetic energy equations thus yield

$$\frac{dE_m}{dt} = -\tilde{I}_p - I'_p\,, \quad (3.7)$$

$$\frac{dE_\ell}{dt} = \tilde{I}_p - I_{\ell t}\,, \quad (3.8)$$

$$\frac{dE_t}{dt} = I'_p + I_{\ell t} - \overline{\phi'}. \quad (3.9)$$

The large-scale structure and fine-grained turbulence production integrals are, respectively,

$$\tilde{I}_p = \int_{-\infty}^{\infty} -\overline{\tilde{u}\tilde{w}}\,\frac{\partial \overline{U}}{\partial z}\,dz \quad (3.10)$$

and

$$I'_p = \int_{-\infty}^{\infty} -\overline{u'w'}\,\frac{\partial \overline{U}}{\partial z}\,dz; \quad (3.11)$$

the large-scale structure and turbulence energy exchange integral is

$$I_{\ell t} = \int_{-\infty}^{\infty} -\left[\overline{\tilde{r}_{xx}\frac{\partial\tilde{u}}{\partial x}} + \overline{\tilde{r}_{xz}\left(\frac{\partial\tilde{u}}{\partial z} + \frac{\partial\tilde{w}}{\partial x}\right)} + \overline{\tilde{r}_{zz}\frac{\partial\tilde{w}}{\partial z}}\right]dz. \quad (3.12)$$

The viscous dissipation integrals in (3.7) and (3.8) have been neglected, while that for the fine-grained turbulence is

$$\overline{\phi'} = \int_{-\infty}^{\infty} \bar{\varepsilon}dz. \quad (3.13)$$

From (3.7)-(3.9) we see that the total energy of the flow, $(E_m+E_\ell+E_t)$, decays according to the rate of viscous dissipation of the "smallest" eddies $\overline{\phi'}$. In the following, we shall discuss the more interesting aspects of the problem via studies of the evolution of the energy contents of the three components of flow.

The Reynolds-average of the numerical calculations reveal that, in spite of local pockets of energy transfer from the fine-grained turbulence to the large-scale structure, the overall direction of energy transfer is from the large to the fine scales and $I_{\ell t} > 0$. The time evolution of this integral is shown in Figure 6a. The large-scale structure production integral $\tilde{I}_p$, shown in Figure 6a, is first positive which accounts for its "amplification"; but subsequently becomes negative, indicating an energy transfer back to the mean motion. In the "inviscid" sense and not surprisingly, this corresponds to a "damped" disturbance. Physically, in the present nonlinear problem, this implies that the large-scale structure has so vigorously modified the mean flow that it has choked off its own energy supply. The initial imbalance between $\tilde{I}_p$ and $I_{\ell t}$ causes the large-scale structure energy to increase at the expense of the mean flow, the subsequent demise of the large-scale structure is due to the energy transferred to the fine-grained turbulence and to the mean motion. The resulting growth and decay of E_ℓ is shown in Figure 6b.

The corresponding mechanisms for the evolution of E_t is shown in Figure 7a. The production of turbulence from the mean flow is enhanced due to the mean flow modification by the large-scale structure. This, and the direct energy transferred from large-scale structure, effects an initial imbalance over $\overline{\phi'}$, E_t evolves from an initial nearly

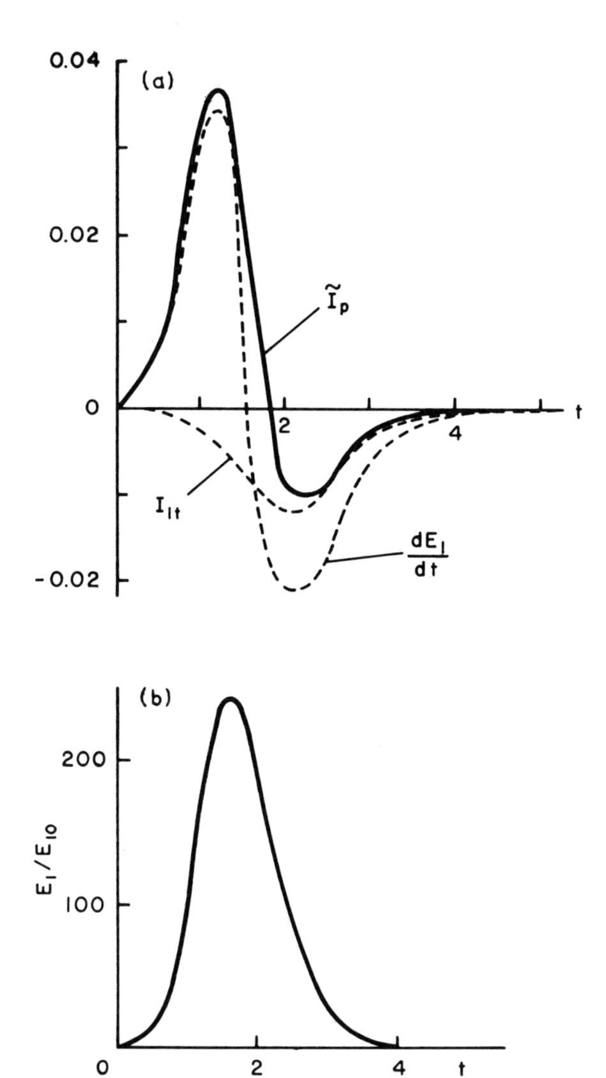

Figure 6.

Evolution of large-scale structure
(a) production and "dissipation" mechanisms and (b) global kinetic energy.

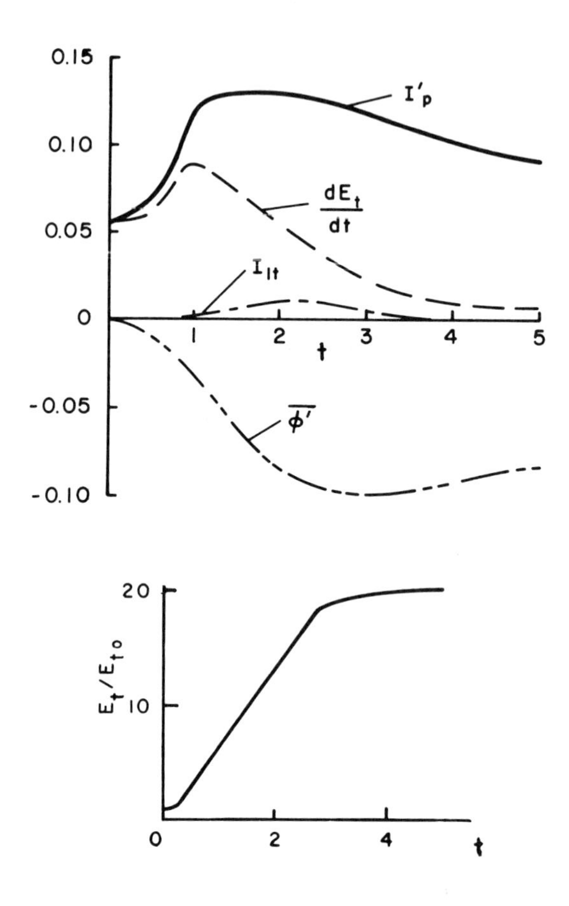

Figure 7.
Evolution of turbulence
(a) production and dissipation mechanisms and (b) global kinetic energy.

self-similar evolution to a new one but at a higher energetic level as shown in Figure 7b. Thus, a "burst" of fine-grained turbulence has occurred at the expense of the large-scale structure. Indeed, this physical picture very much resembles that of Favre-Marinet and Binder's [15] well-controlled experiments illustrated in Figure 1 if we compare E_ℓ and E_t, which, in fact, are amplitudes of the large-scale structure and the fine-grained turbulence, with their respective measurements. Here, however, we have attempted to explain the the mechanisms leading to the evolution of the nonlinear interactions. We make the observation here that, in general, the energy transfer integrals $\tilde{I}_p$ and $I_{\ell t}$ in (3.8) are not necessarily in balance and are strongly evolving in time for the free shear flow problem. In this case, marginal stability ideas would not hold. There are rather strong nonlinear complings between the three components of flow and the large-scale structure, unless exceptionally "weak" cannot be considered independently of either the developing mean flow or of the fine-grained turbulence.

The growth and decay of the large-scale structure energy content E_ℓ is similar to the height of the closed streamline of the conditionally averaged flow denoted by H. This quantity, normalized by its initial value H_0, is shown in Figure 8. The mean shear layer thickness

$$\delta(t) = \left[\int z^2 \partial\bar{U}/\partial z \, dz \Big/ \int \partial\bar{U}/\partial z \, dz\right]^{1/2},$$

normalized also by its initial value, is also shown in Figure 8. Its initial growth is in a self-similar manner $\delta \sim t$ but is modified by the nonequilibrium, nonlinear interactions. The dissipation length scale of the fine-grained turbulence L_ε is estimated by a local equilibrium argument which leads to $L_\varepsilon \approx E_t^{3/2}/\overline{\phi'}$. It is normalized by its initial value $L_{\varepsilon 0}$ and shown in Figure 8 also. Accompanying the vigorous activities of the large-scale structure is a rapid "burst" of finer scales of the turbulence. However, as the large-scale structure eventually decays, the scale of the fine-grained turbulence increases in order to maintain the growth of the mean shear layer. In fact, as shown in

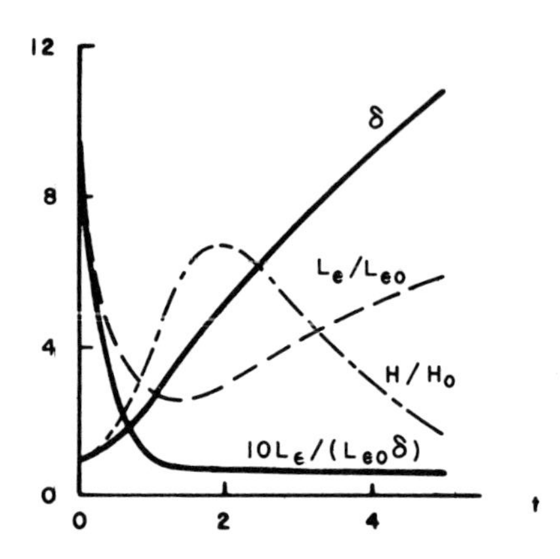

Figure 8.
Evolution of shear layer thickeness, large-scale structure "height" and turbulence dissipation length scale.

Figure 8, eventually L_ε scales like δ. A careful examination of the structure in Brown and Roshko's [8] optical observation reveals a similar enlargement of the graininess of the fine-grained turbulence as the flow spreads downstream, accompanied by the demise of the large-scale structure.

4. THE APPROXIMATE DESCRIPTION OF THE ROUND, TURBULENT JET PROBLEM

Although we have derived much physical information from the simple shear layer calculations of Section 3, the numerical work involved makes it rather difficult to study the important effects of spectral behavior and of the initial conditions. However, the results of Section 3 can be used as a guide to obtain an approximate description of the interaction problem from which spectral effects and initial conditions can be more easily studied. The approximate descriptions of the interactions between large-scale structure and fine-grained turbulence plane turbulent shear layers were given previously in [37] and [1]. Here, we shall discuss the technologically important round jet problem [40] and bring out the important effects, which are not merely novel, when we take the presence of fine-grained turbulence into serious account.

The basis for the study is the radially integrated, time-averaged kinetic energy equations for the mean flow, large-scale structure and fine-grained turbulence, similar to (3.7)-(3.9) but modified here to the spatially developing problem and to account for the geometrical effects pecular to the round jet. The use of the radial, or in general, cross-stream integrated form of the conservation equations is not unlike the "shallow water" approximation [62]. Although it is more involved when we deal with the large-scale structure from an instability point of view in that suitable spectral properties have to be properly taken into account.

However, there has been some success in applying such techniques to the nonlinear instability stages of laminar free shear flow transition when compared with observations [27], [35]. The attractiveness is that the relevant physical information can be extracted from the problem with relative ease, particularly in the absence of a small parameter and when direct numerical computations are difficult and not preferred at the outset.

Accompanying the radially integrated conservations equations are the necessary shape assumptions for the radial structure of the flow components which will reduce the problem to streamwise-dependent parameters. These then must be solved from the streamwise nonlinear, nonequilibrium evolution problem. For the mean flow and the fine-grained turbulence stresses the shape assumptions are entirely similar to von Kármán's integral technique. In this case, the mean velocity distribution would be taken to be of the form $\overline{U} = \overline{U}_c f(r;\theta)$, where r is the radial distance, $\theta(x)$ is a streamwise or x-dependent length scale characterizing the radial spreading of the mean flow and $\overline{U}_c$ is the jet centerline velocity. However, $\overline{U}_c = \overline{U}_c(\theta)$ from momentum considerations. In this case $\theta(x)$ is the only parameter that characterizes the mean flow. The mean turbulence stress are expressed as $\overline{u_i'u_j'} = E(x)g_{ij}(r;\theta)$, where the shape functions g_{ij} are guided by observations and are so normalized so as to make $E(x)$ the yet to be solved energy density of the fine-grained turbulence over a local cross-section of the jet. The large-scale structure shape assumption follows ideas on finite-amplitude disturbances in parallel laminar transitional flows due to Stuart [54]. Extensions to developing mean flows [27], [36] would then involve the interpretation of the radial or cross-stream shape as being given by a local linearized theory with the unknown amplitude function $A(x)$ to be solved simultaneously with $E(x)$ and $\theta(x)$. In the presence of fine-grained turbulence, however, the excess turbulence stresses $\tilde{r}_{ij}$ are coupled to the large-scale structure flow quantities $\tilde{q}$ even in a local linear theory as shown by Elswick [14] and Reynolds [48] for the plane problem. A consistent shape assumption for both $\tilde{q}$ and $\tilde{r}_{ij}$ would appear as

$$\begin{bmatrix} \tilde{q}(x,r,\phi,t) \\ \tilde{r}_{ij}(x,r,\phi,t) \end{bmatrix} = A(x)\exp i\left[\int_0^x \alpha_r(\xi)\,d\xi - \omega t + n\phi\right]\begin{bmatrix} \hat{q}(r;\theta,St,n) \\ E(x)\hat{r}_{ij}(r;\theta,St,n) \end{bmatrix} + \text{complex conj.}, \tag{4.1}$$

where $A(x)$ is the complex amplitude function, x,r,ϕ are the streamwise, radial and azimuthal coordinates, α_r is the streamwise wave number ($\int_0^x \alpha_r d\xi$ is Gaster's [18] wave number), n is the azimuthal wave number, ω is the frequency, $St = \omega d/2\pi\bar{U}_e$ is the Strouhal number, $\bar{U}_e$ is the jet exit velocity and d the jet exit nozzle diameter. The radial shape functions are denoted by $(\hat{\ })$ and are implicitly dependent upon x through their dependence on the local mean flow length scale $\theta(x)$. The structure of the transport equations for $\tilde{r}_{ij}$ shows that its primary sources are the advection of the mean stresses by the large-scale structure and the work done by the mean stresses against the large-scale structure rates of strain. In both cases and to lowest order $\tilde{r}_{ij} \sim A(x)E(x)$. Thus anticipating that the energy transfer between large scale structure and the fine-grained turbulence would be scaled according to $|A(x)|^2E(x)$. The eigenfunctions $\hat{q}$ are so normalized as to make $|A(x)|^2$ the kinetic energy content of the large-scale structure over a cross-section of the jet, similar to the interpretation attributed to $E(x)$.

The shape functions $\hat{q}$ and $\hat{r}_{ij}$, unlike those for the mean flow and mean turbulent stresses, are here generated by solving a local eigenvalue problem for each local value of $\theta(x)$. Some simplifications are possible and this relies on the dynamical nature of the instabilities associated with inflexional mean flows. In this case, the instability characteristics of the large-scale structure are primarily determined by the "inviscid" instability problem obtained by dropping all "diffusive" effects such as those due to $\hat{r}_{ij}$. This then uncouples the $\hat{q}$ and $\hat{r}_{ij}$, as argued by Liu and Merkine [37]. In this case, the inviscid $\hat{q}$ is then used to solve for $\hat{r}_{ij}$. Although approximate, this nevertheless gives the proper spectral (i.e., St) dependence of the problem and from which the energy transfer between the large-scale structure and fine-grained turbulence can be approximately but rationally obtained.

The resulting nonlinear interaction problem is obtained from the integrated kinetic energy equations of the mean flow, large-scale structure and fine-grained turbulence,

$$\frac{1}{2}\frac{d}{dx} I_1(\theta) = -I'_{Rs}(\theta)E - \tilde{I}_{Rs}(\theta;St,n)|A|^2, \tag{4.2}$$

$$\frac{d}{dx}[I_2(\theta;St,n)|A|^2] = \tilde{I}_{Rs}(\theta;St,n)|A|^2 - I_{wt}(\theta;St,n)|A|^2E, \tag{4.3}$$

$$\frac{d}{dx}[I_3(\theta)E] = I'_{Rs}(\theta)E + I_{wt}(\theta;St,n)|A|^2E - I_\varepsilon(\theta)E^{\frac{3}{2}}. \tag{4.4}$$

The initial conditions are $\theta(0) = \theta_0$, $|A(0)|^2 = |A|_0^2$ and $E(0) = E_0$; we need also to specify St and n. The physical mechanisms represented by the right side of (4.2)-(4.4) are identical to those of (3.7)-(3.9). Here the left sides of (4.2)-(4.4) represent the streamwise flux of kinetic energy of the respective components. In the mean flow energy equation (4.2) $I_1(\theta)$ represents the mean flow energy advection integral and $dI_1(\theta)/d\theta < 0$. I'_{Rs} and $\tilde{I}_{Rs}(\theta;St,n)$ are the energy production integrals of the fine-grained turbulence and large-scale structure, respectively. According to (4.2), as long as energy is extracted from the mean flow $d\theta/dx > 0$ and the mean flow will spread downstream. It is obvious that the mean flow spreading rate is dictated by both the large-scale structure and the fine-grained turbulence. A return of energy back to the mean flow, such as via the "damped" disturbance interpretation of Section 3, would contribute to a decrease in the rate of spread as has been found experimentally by Durgin and Karlsson [13]. In (4.3) and (4.4), $I_2(\theta;St,n)$ and $I_3(\theta)$ are the advection integrals of the large-scale structure and fine-grained turbulence, respectively. $I_{wt}(\theta;St,n)$ is the integral representing the energy transfer between the large-scale structure and fine-grained turbulence, the integrand of which has been the subject of approximate calculations from conservation equaitons. Lastly, $I_\varepsilon(\theta)$ is the viscous dissipation integral of the fine-grained turbulence. Integrals which arise from the mean flow and the mean turbulence stresses are functions of θ alone, whereas those from the large-scale structure depends not only on θ, but also on the frequency, or more precisely, the Strouhal number. For instance, within certain bands of frequency range the large-scale structure is more efficient in extracting energy from the mean flow. Such integrals also depend on the azimuthal wave number n, where $n = 0$ is the axially symmetric mode and $n = 1$ the first helical mode (see, for instance, Batchelor and Gill [4]).

Some of the closure problems have already been circumvented within the present framework. The triple correlations do not appear in the integrated energy equation. Being of higher order, they would not appear in the linearized eigenvalue problem for the shape functions for the large-scale structure; however, the pressure-velocity strain closure and the simple closure for the viscous dissipation rate have been applied. The latter is evident in the integral energy equation for the fine-grained turbulence where the viscous dissipation rate is proportional to $E^{3/2}$.

We shall compare the results of the present much simplified framework with the observations of Favre-Marinet and Binder [15]. Their jet centerline measurements, shown in Figure 1, were normalized by the measured mean velocity on the centerline. In the comparison, it is preferred to normalize the root mean square (r.m.s.) fluctuation velocities by the constant jet exit velocity. Figure 9 shows the

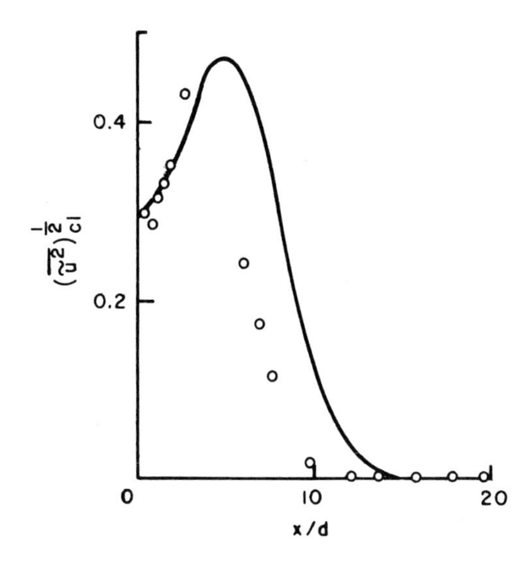

Figure 9.
Large-scale structure r.m.s. axial velocity along the jet centerline. Comparison with Favre-Marinet and Binder's experiments [15], St = 0.18.

comparison of the r.m.s. large-scale structure streamwise velocity component along the jet centerline for St = 0.18 and $n = 0$. The initial forcing velocity is relatively large (≈ 30%) and thus the comparison here places a severe test upon the present framework. The r.m.s. fine-grained turbulence streamwise velocity component along the jet is shown in Figure 10. The St = 0.18 case shows that the fine-grained turbulence is enhanced compared to the St = 0 (unforced) case. Both $\left(\overline{\tilde{u}^2}\right)^{1/2}$ and $\left(\overline{u'^2}\right)^{1/2}$ were given in the experiment at the nozzle exit, thus $|A|_0^2$ and E_0 were fixed in the calculations. However, θ_0 was not explicitly given in the experiments but has been estimated to

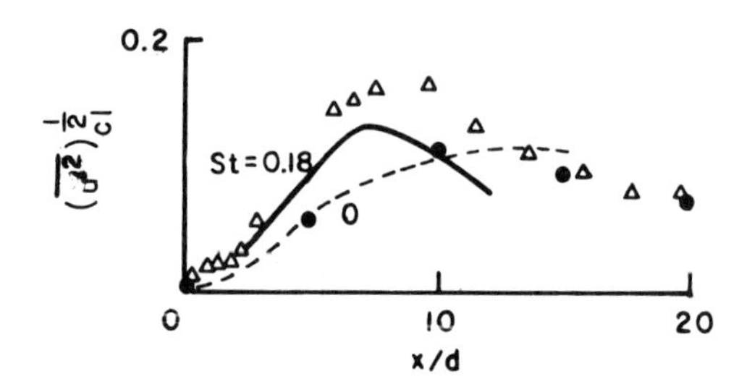

Figure 10.
Fine-grained turbulence r.m.s. axial velocity along the jet centerline. Comparison with Favre-Marinet and Binder's with experiments [15] for St = 0.18(Δ) and for the unforced jet St = 0(o).

be about $\theta_0 \approx 0.3$ because of the large amplitude forcing. The dominant features of the observations are explanable from the calculations, although the details leave much to be desired. It is also possible to obtain certain detailed structural comparisons in terms of the streamwise evolution of the radial distribution of both $\left(\overline{\tilde{u}^2}\right)^{1/2}$ and $\left(\overline{u'^2}\right)^{1/2}$; such comparisons have been rather encouraging [40] inspite of the fact that the present considerations are not meant to yield structural details. The dominant physical mechanisms leading to the growth and decay of the large-scale structure, the "burst" of the fine-grained turbulence and the spreading of the mean flow, according to (4.2)-(4.4), are very much the same as those discussed in Seciton 3.

In the experiments of Favre-Marinet and Binder [5,15], which corresponds to rather strong forcing, the potential core of the jet is very short and the jet exit mean velocity profile is much fuller than a "top hat" profile. The experiments done by Moore [43] included cases of "weak" forcing and the jet exit velocity is very nearly a "top hat" profile. Like the experiments of Crow and Champagne [12], the signals were filtered (rather than conditionally averaged) at the large-scale structure frequency. Under the premise that the fine-grained turbulence energy at the filtered frequency is small compared to that of the monochromatic large-scale structure, comparisons between the filtered data and calculated large-scale structure quantities could still be made. Such comparisons, however, are not entirely free from such difficulties. However, Moore [43] obtained the response of the jet to a wide spectrum of excitation frequencies and some comparisons with such results would be helpful in assessing the spectral effects obtained from the present much simplified framework. In the case of weak excitation, Moore's [43] initial conditions correspond to the present $|A|_0^2 \approx 3.2 \times 10^{-7}$, $E_0 \approx 7.8 \times 10^{-4}$ and $\theta_0 \approx 0.034$. Both the r.m.s. velocity and pressure were measured along the jet centerline and the streamwise evolution for each frequency component agree with the present calculations reasonably well. We refer to [40] for further details. The growth and decay of the large-scale structure results in a peak for each Strouhal number. Figure 11 shows the comparison between the

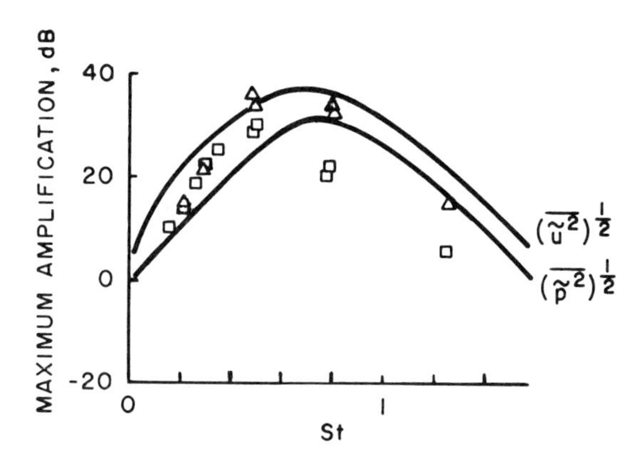

Figure 11.
Maximum large-scale structure amplification as a function of Strouhal number; comparison with Moore's experiments [43].

present calculations for the maximum amplification of both $\left[\overline{\tilde{u}^2}\right]^{1/2}$ and $\left[\overline{\tilde{p}^2}\right]^{1/2}$ as functions of the Strouhal number and Moore's [43] measurements. For this set of initial conditions, the peak amplification is about $St \approx 0.6 - 0.7$. In Figure 12, we show the comparison of the phase velocity C_{ph} as a function of the large-scale structure frequency fx/d for two different streamwise sections along the jet and the agreement is again reasonably good. In the present formulation the phase velocity is obtained from a local linearized theory, as a function of $\theta(x)$ and the nonlinear analysis (4.2)-(4.4) then furnishes the distribution of $\theta(x)$ and hence C_{ph}, along the jet. All the comparisons up to now are for the $n = 0$ mode. Some confidence has now been gained with the formulation so that we could be in a position to discuss and try to understand those physical situations where quantitative observations are not as yet available or are difficult to obtain.

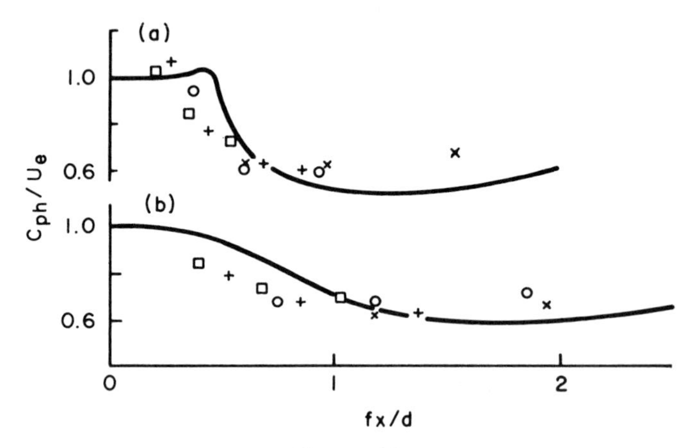

Figure 12.
Large-scale structure phase velocity as a function of normalized frequency; comparison with Moore's experiments [43], jet Mach number M = 0.30, ×; M = 0.49, o; M = 0.66, +; M = 0.83, □.
(a) x/d = 1.19, (b) x/d = 2.38.

It is well known from the work of Batchelor and Gill [4] on the dynamical instability of round jets that the $n = 1$ helical mode is more unstable than the $n = 0$ mode. In the absence of fine-grained turbulence, the production mechanism alone determines the development of the large-scale structure. The competition of the "turbulence dissipation" mechanism enters into the energy balance in a very important way in determining the survival of the large-scale structure. For the $n = 0$ axially symmetric large-scale structure in a turbulent jet, its energy exchange with the fine-grained turbulence is accomplished by the work done by the normal stresses $\tilde{r}_{xx}, \tilde{r}_{rr}, \tilde{r}_{\phi\phi}$ and the single shear stress $\tilde{r}_{xr}$

$$\overline{\tilde{r}_{xx}\frac{\partial\tilde{u}}{\partial x}},\ \overline{\tilde{r}_{rr}\frac{\partial\tilde{u}}{\partial r}},\ \overline{r_{\phi\phi}\frac{\tilde{v}}{r}},\ \overline{\tilde{r}_{xr}\left(\frac{\partial\tilde{v}}{\partial x}+\frac{\partial\tilde{u}}{\partial r}\right)}.$$

For the $n = 1$ helical mode, additional shear stresses are set-up due to the azimuthal transport of axial momentum $\tilde{r}_{x\phi}$

and of radial momentum $\tilde{r}_{r\phi}$ which results in the additional energy transfer mechanisms

$$\overline{\tilde{r}_{x\phi}\left(\frac{\partial\tilde{u}}{r\partial\phi} + \frac{\partial\tilde{w}}{\partial x}\right)}, \quad \overline{\tilde{r}_{r\phi}\left(\frac{\partial\tilde{v}}{r\partial\phi} + \frac{\partial\tilde{w}}{\partial r} - \frac{\tilde{w}}{r}\right)}.$$

The presence of the azimuthal velocity $\tilde{w}$ and the rate of change in the azimuthal direction also set up an additional normal stress mechanism

$$\overline{\tilde{r}_{\phi\phi}\frac{\partial\tilde{w}}{r\partial\phi}}$$

for the helical mode. In Figure 13 we show the radial distribution energy exchange mechanisms, obtained in the

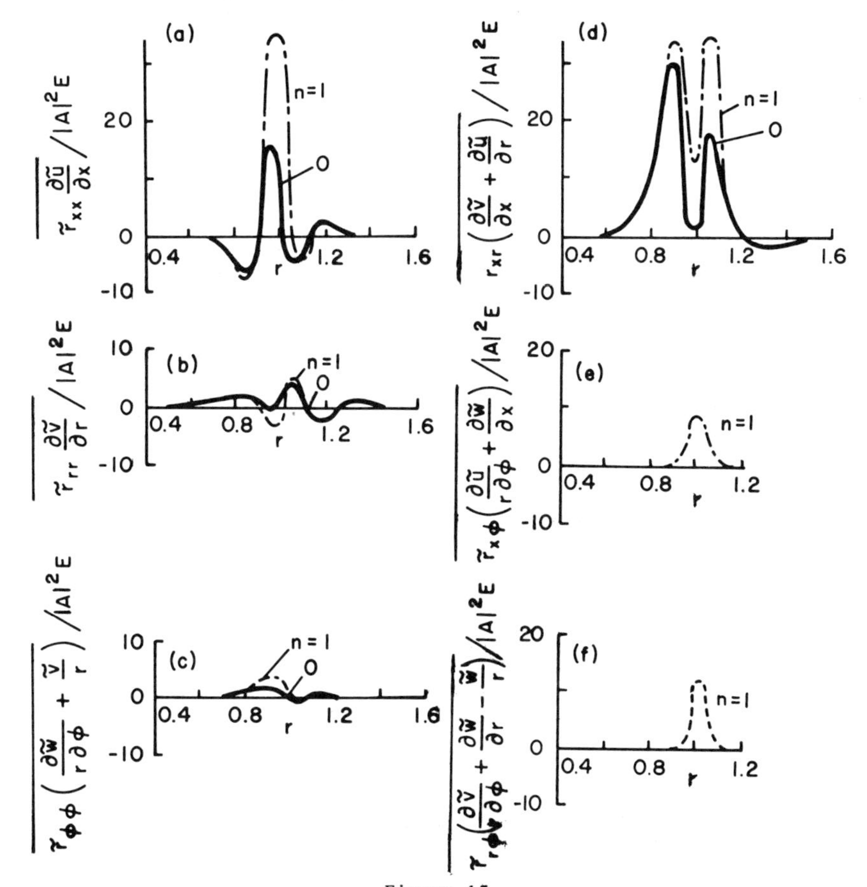

Figure 13.
Mechanisms for energy transfer between large-scale structure and fine-grained turbulence; St = 0.5, θ = 0.08.

approximate manner as previously discussed, for St = 0.5 and a local $\theta = 0.08$. The overall energy transfer I_{wt} is obtained via a radial integration. In this case, the xr-mechanism is largely positive indicating an energy transfer to the fine-grained turbulence with the $n = 1$ mode having the larger contribution. The $n = 1$ mode also dominates the xx-mechanism, while that of $n = 0$ is partially cancelled by negative values of the energy transfer. The rr- and $\phi\phi$-mechanisms are ineffective in that they are either small or are self-cancelling upon radial integration, however, it is noticed here that the $n = 1$ mode dominates the $\phi\phi$-mechanism. The $x\phi$- and $r\phi$-mechanisms are both positive and are additional "dissipative" mechanisms for the $n = 1$ mode. While the production mechanism may be stronger for the $n = 1$ mode compared to that for the $n = 0$ mode at the same Strouhal number and same initial conditions, the $n = 1$ mode has an even stronger turbulence dissipation mechanism which would lead to its early demise following a more vigorous growth. To illustrate this point, we show in Figure 14 the

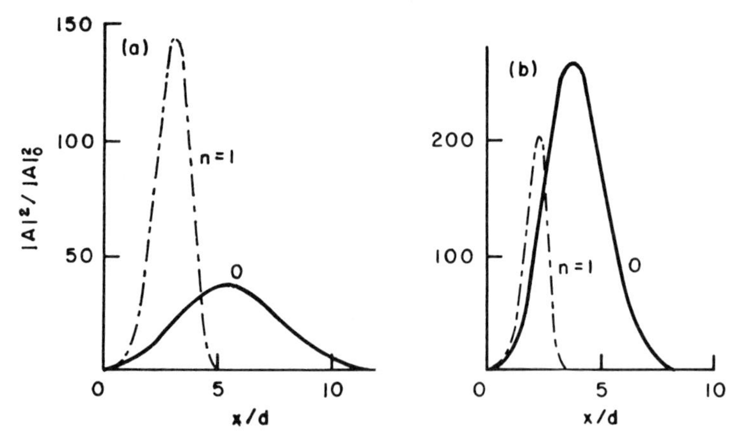

Figure 14.
Large-scale structure kinetic energy ratio along the jet: comparison between the n = 0 and n = 1 modes. (a) St = 0.35, (b) St = 0.50.

development of the large-scale structure energy ratio $|A|^2/|A|_0^2$ for St = 0.35 and 0.50 for $|A|_0^2 = 10^{-5}$, $E_0 = 10^{-3}$ and $\theta_0 = 0.034$. The comparison between different Strouhal numbers indicate that the problem is strongly frequency dependent. This is because the production mechanism is strongly frequency dependent as shown in Figure 15.

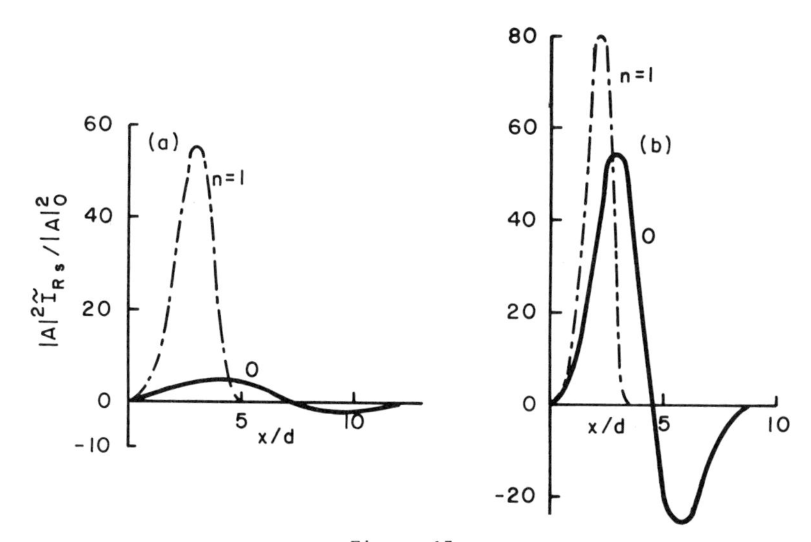

Figure 15.
Large-scale structure production rate along the jet: comparison between the n = 0 and n = 1 modes. (a) St = 0.35, (b) St = 0.50.

The reversal of the large-scale structure Reynolds stress, resulting in energy transfer back to the mean flow is associated more with the $n = 0$ mode. The dissipative mechanism is shown in Figure 16, which is rather spectacular for the $n = 1$ mode. In general, lower Strouhal number large-scale structures have longer streamwise lifespan compared to the higher frequency components as is found here and elsewhere [32,33].

We have shown that the Strouhal number and the azimuthal wave number play an essential role in the evolution of the large-scale structure. In addition, the initial conditions such as $|A|_0^2$, E_0 and θ_0 also play an important role. In Figure 17, we show the effect of initial turbulence level E_0 on the downstream development of the flow quantities. Figure 17a shows the large-scale structure for $St = 0.35$ and $|A|_0^2 = 10^{-5}$, and to fix our ideas, for $n = 0$. If the turbulence is initially weak, the large-scale structure energy

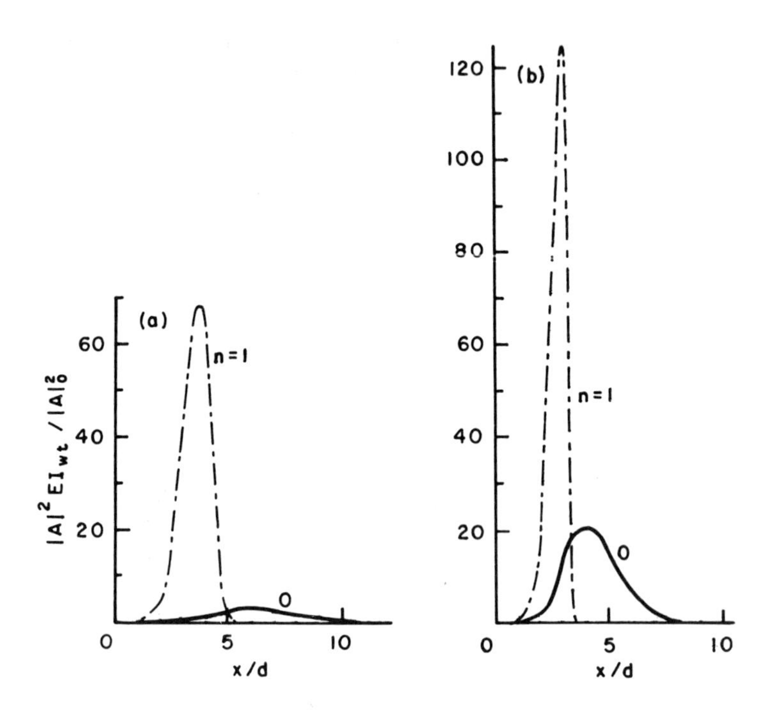

Figure 16.
Large-scale structure to fine-grained turbulence energy transfer rate along the jet: comparison between the n = 0 and n = 1 modes. (a) St = 0.35, (b) St = 0.50.

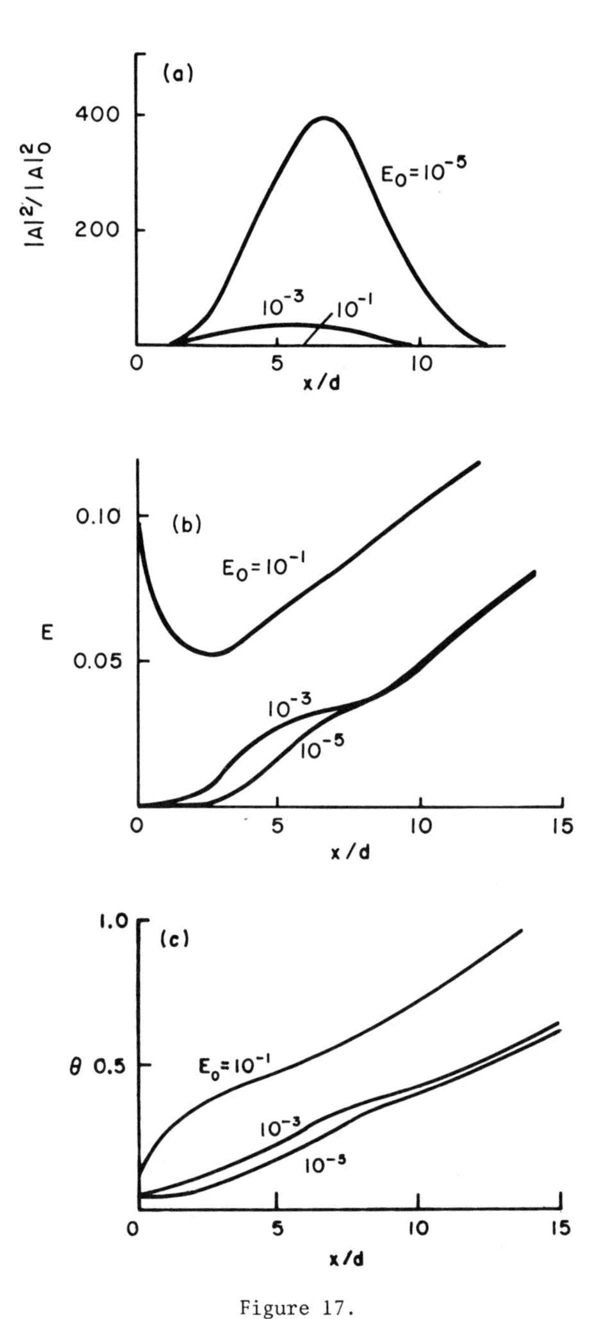

Figure 17.
Effect of initial fine-grained turbulence levels on the evolution of (a) large-scale structure energy ratio, (b) fine-grained turbulence energy, (c) mean flow length scale. St = 0.35, n = 0.

ratio $|A|^2/|A|_0^2$ attains a much larger amplification. For sufficiently strong turbulence levels the large-scale structure does not develop at all. This is entirely consistent with the observations of Chandrsuda, Mehta, Weir and Bradshaw [10]. The development of fine-grained turbulence energy is shown in Figure 17b. The large dip for the $E_0 = 10^{-1}$ case is due to the relatively large viscous dissipation ($\propto E^{3/2}$) relative to production ($\propto E$). Eventually, E becomes self-similar and develops like x, however, the relative self-similar level for the $E_0 = 10^{-1}$ case remains higher. Figure 17c shows the corresponding mean flow length scale development. The fine-grained turbulence production rate increases with E_0 resulting in an increase in the jet spreading. That the mixing rate can also be enhanced by the fine-grained turbulence is indicated by the experiments of Hussain and Zedan [22]. We can assess the relative effectiveness of the large-scale structure and fine-grained turbulence in enhancing the <u>overall</u> mixing process from the results obtained thus far. If we take $\theta \approx 0.25$-0.30 as value indicating that the potential core of the jet is very nearly terminated, then at $St = 0.35$ an increase of $|A|_0^2$ from 10^{-5} to 10^{-3} produces a 28% reduction in the axial length of the potential core while increasing E_0 from 10^{-5} to 10^{-3} produces only an 18% reduction. In this case, at least, the large-scale structure appears to be relatively more efficient as a means to enhance jet spreading as already indicated by the experiments of Binder and Favre-Marinet [5].

The initial levels of the large-scale structure is important to the nonequilibrium evolution in two ways. First, if it is sufficiently large, it would distort the mean flow to such an extent that it chokes off its own energy supply; second, a large initial level also implies that the turbulent dissipation mechanism, being proportional to $|A|^2E$, would be strong. In this case, if all else is fixed large, initial $|A|_0^2$ is associated with lesser relative amplification and a shorter streamwise lifespan. This is illustrated in Figure 18 in terms of the evolution of $\left(\overline{\tilde{u}^2}\right)^{1/2}$ on the jet centerline. The evolution of the $\left(\overline{u'^2}\right)^{1/2}$.

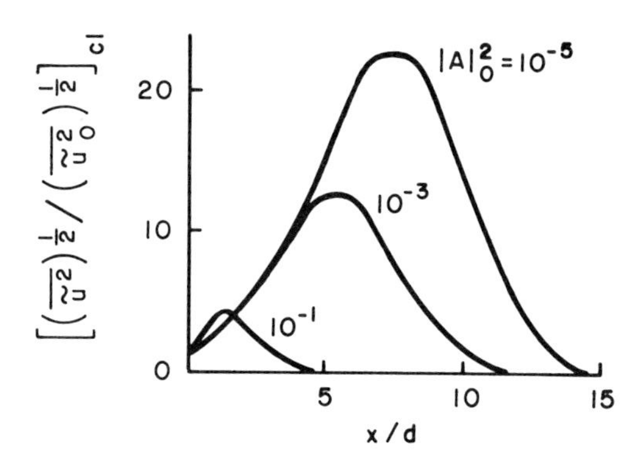

Figure 18.
Large-scale structure r.m.s. streamwise velocity component ratio along the jet centerline for several initial large-scale structure energy levels. St = 0.35, n = 0.

level is shown in Figure 19. The larger $|A|_0^2$ cases provide more sources of energy for the fine-grained turbulence. However, in this case, the potential core is terminated rapidly

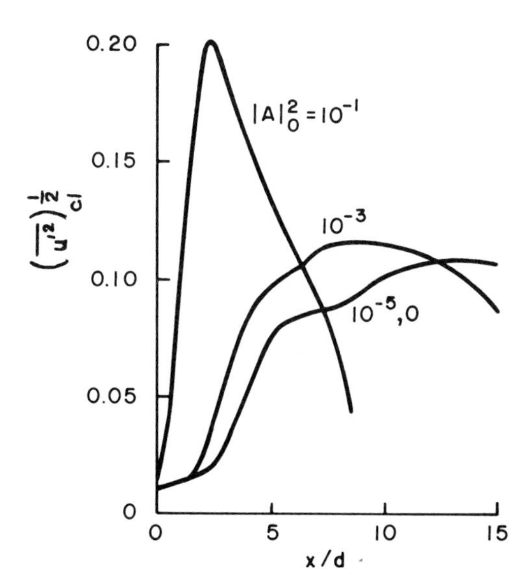

Figure 19.
Fine-grained turbulence r.m.s. axial velocity component along the jet centerline for several-initial large-scale structure energy levels. St = 0.35, n = 0.

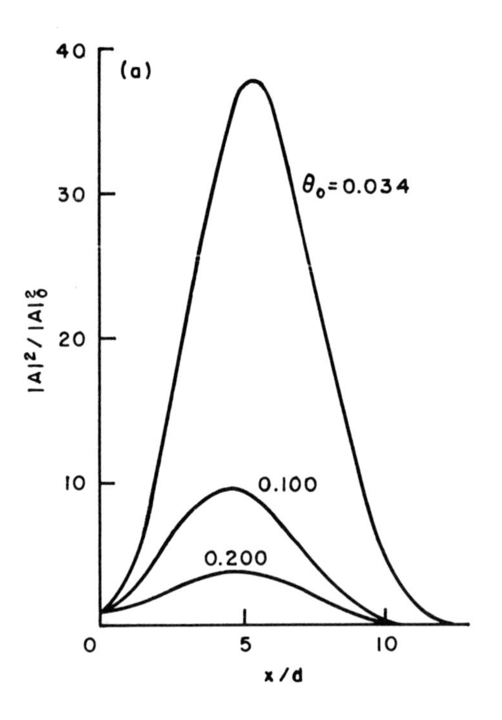

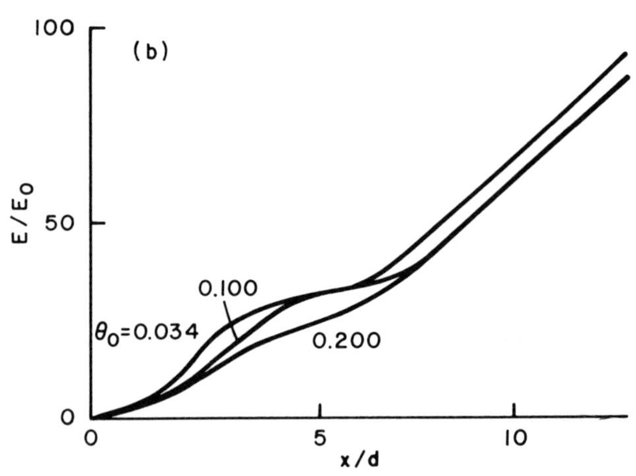

Figure 20.
The effect of initial momentum thickness on the evolution of (a) large-scale structure energy ratio, (b) fine-grained turbulence energy ratio, (c) mean flow length scale.
$St = 0.35$, $n = 0$, $|A|_0^2 = 10^{-5}$, $E_0 = 10^{-3}$.

as a result of the mixing promoted by both components of the fluctuations.

The initial mean velocity profile, characterized by θ_0, has a strong influence on the subsequent development of the nonlinear interactions through the production mechanisms for both the large-scale structure and the fine-grained turbulence. Thus, smaller values of θ_0, corresponding to "top hat" mean velocity profiles, provide stronger production mechanisms, while larger θ_0 corresponds to more developed initial mean velocity profiles. Figure 20a shows that the large-scale structure energy ratio which can be suppressed by smoothing the initial mean velocity profile. This is in full agreement with the observations of Chen and Templin [9]. The corresponding evolution of the turbulence energy is shown in Figure 20b. Although E/E_0 reaches a self-similar development far downstream, the relative level for the $\theta_0 = 0.034$ case is higher owing to the energy transferred from the strongly developed large-scale structure. The streamwise development of the mean flow length scale θ is shown in Figure 20c. Far downstream the rate of spread becomes

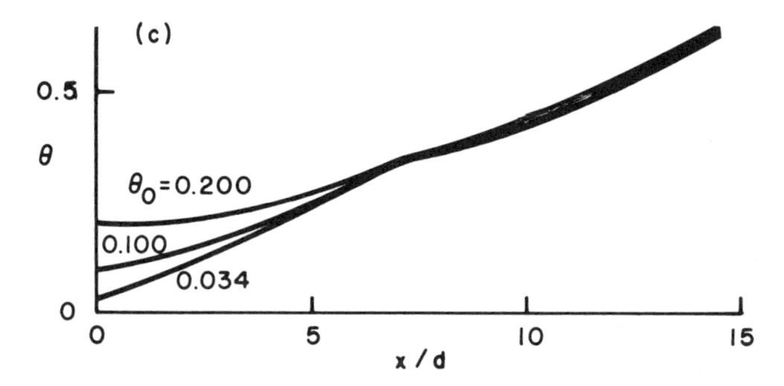

Figure 20(c).
For description, see under Figure 20(b).

independent of θ_0, which is in accordance with Hussain and Zedan's [23] observations. However, this illustrates that the far downstream mean flow spreading rate gives very little or no information about the nonequilibrium region of interest.

5. AGGLOMERATION OF LARGE-SCALE STRUCTURES AND SUBHARMONIC FORMATION

The discussion as to the role of agglomeration of neighboring large-scale structures have been given recent impetus [50,63]. This is generally associated with subharmonic formation. Sato [51] observed what was then the rather curious development of a subharmonic disturbance in the nonlinear stages of a separated shear layer in the transition from laminar to turbulent flow. On the other hand, such subharmonic disturbances were not observed in the transition region in the wake behind a flat plate [52]. In both cases, mean flow spreading took place under the vigorous momentum transport activities of the developing nonlinear disturbances. From the kinetic energy considerations of the last Section, we can understand that mean flow spreading takes place when energy is fed to the disturbing motion on the average. If such spreading takes place in the presence of vigorous subharmonic activities it necessarily implies that the subharmonics are extracting energy from the mean motion. On the other hand, the detection of spreading itself does not necessarily imply the presence of the formation of subharmonics. More visually, the agglomeration or "pairing" of large-scale structures, is now rather well recognized as occurring in the transition region of laminar free shear layers [7,16,41,42]. More recently, it is shown that such agglomeration events can be well controlled including their suppression [6,20].

A theoretical explanation for the presence of a subharmonic component in free shear layers was given by Kelly [24] some time ago. The stability of the combined mean flow and nearly equilibrated finite, fundamental disturbance was studied. The conditions that favor the subharmonic formation are that the finite fundamental disturbance must reach a sufficiently large threshold and that the phase relationship between the fundamental and the subharmonic perturbation must be "favorable". Kelly's mechanism for subharmonic formation has been confirmed numerically [46] (although the numerical computations were for a stratified shear layer). The physical interpretation of Kelly's mechanism can be obtained again from the kinetic energy considerations.

To this end, it is appropriate to refer to the kinetic energy equation obtained by Stuart [55] for the temporal disturbance of odd multiples of the wave number $(2n+1)\alpha$ (with velocities denoted by $\tilde{u},\tilde{w}$) interacting with disturbances of even multiples of the wave number $2n\alpha$ (with velocities denoted by $\tilde{\tilde{u}},\tilde{\tilde{w}}$), in a horizontally homogeneous mean shear flow,

$$\frac{1}{2}\frac{d}{dt}\int \overline{(\tilde{u}^2+\tilde{w}^2)}\,dz = \int -\overline{\tilde{u}\tilde{w}}\,\frac{\partial\bar{U}}{\partial z}\,dz + \int \left[\overline{-\tilde{u}^2\,\frac{\partial\tilde{\tilde{u}}}{\partial x}} - \overline{\tilde{u}\tilde{w}\left(\frac{\partial\tilde{\tilde{u}}}{\partial z}+\frac{\partial\tilde{\tilde{w}}}{\partial x}\right)} - \overline{\tilde{w}^2\,\frac{\partial\tilde{\tilde{w}}}{\partial z}}\right]dz - \overline{\tilde{\phi}}. \tag{5.1}$$

The over-bar denotes the Reynolds average over the largest horizontal periodicity, which is here $2\pi/\alpha$. Although Stuart [55] originally intended the derivation of (5.1) with $\tilde{u}$ as the fundamental component and $\tilde{\tilde{u}}$ as the first harmonic, however, (5.1) can be reinterpreted with $\tilde{u}$ now as the subharmonic with wavelength $2\pi/\alpha$ and $\tilde{\tilde{u}}$ as the fundamental with wavelength π/α. Here $\overline{\tilde{\phi}}$ would be interpreted as the "dissipation" rate. In the Stuart-Kelly problems $\overline{\tilde{\phi}}$ would indeed be the viscous dissipation rate. To interpret (5.1) in a much broader sense, $\overline{\tilde{\phi}}$ would include the fine-grained turbulence dissipation. Equation (5.1) also holds for the spreading shear layer. In fact, (5.1) is quite similar in form to the energy balance of the large-scale structure in Section 3 (or Section 4), where we discussed the energy transfer between two disparate scales of oscillations: large-scale periodic and fine-grained random. Here we shall concentrate on discussing the energy transfer between two periodic oscillations of different wavelengths. The energy transfer in both situations are due to the stresses doing work against the appropriate rates of strain.

The subharmonic component has its own production mechanism corresponding to the first term on the right of (5.1). The second term on the right is the possible energy source from the fundamental component. In order for the fundamental to play a competitive role in the supply of energy then clearly $\tilde{\tilde{u}}$ (and $\tilde{\tilde{w}}$) must be sufficiently large or that its associated rates of strain be sufficiently large. The discussion of the magnitude is clearly insufficient. The relative phases between the subharmonic stresses $-\tilde{u}^2$, $-\tilde{u}\tilde{w}$ and $-\tilde{w}^2$ and the corresponding rates of strain of the fundamental are crucial in determining the direction of energy transfer. An insight into the role of such magnitude and crucial phase relationships can be gained from the discussions of the similar mechanisms in the large-scale structure and fine-grained turbulence problem [1,19,37,40]. Although in the derivation of Kelly's mechanism [24] use was made of "weakly nonlinear" disturbances, the general physical consequences are much more far-reaching, particularly in the light of observations based on the energy considerations (5.1) which holds for strongly amplified disturbances as well.

The possible enhancement of fine-grained turbulence in an agglormeration event can be studied approximately through a more detailed interpretation of $\tilde{\phi}$ in (5.1), with coupling to the energy balance equations of the fundamental component, the mean flow and the fine-grained turbulence. A direct numerical integration approach, similar to the framework described in Section 3, would involve extending the horizontal integration domain to two wave lengths of the fundamental component with initial conditions including the subharmonic perturbation. This has recently been done by Knight [26] on the basis of a different Reynolds stress closure than that used in Section 3. However, Knight's [26] particular numerical results indicate that the fine-grained turbulence is feeding energy, on the average, to the large-scale structures during the agglomeration process; the subsequent decay of the large-scale structure is due primarily to a return of its energy back to the mean flow. The "anti-cascading" process of the fine-grained turbulence from Knight's result somewhat resembles that of the simulation of "turbulence" by a cloud of two-dimensional potential vortices [3].* However, the

*We refer to Aref and Siggia [3] for a review of the literature on the "atomistic" simulation of turbulence by clouds of point vorticies.

closure method used [26] depicts three-dimensional, dissipative turbulence. The interaction between Kelly's mechanism and fine-grained turbulence still poses a rather intriguing problem theoretically and experimentally.

REFERENCES

1. Alper, A. and J. T. C. Liu, On the interactions between large-scale structure and fine-grained turbulence in a free shear flow. II. The development of spatial interactions in the mean. Proc. R. Soc. Lond. A 359 (1978), 497-523.
2. Amsden, A. A. and F. H. Harlow, Slip instability, Phys. Fluids 7 (1964), 327-334.
3. Aref, H. and E. D. Siggia, Vortex dynamics of the two-dimensional turbulent shear layer, J. Fluid Mech. 100 (1980), 705-737.
4. Batchelor, G. K. and A. E. Gill, Analysis of the stability of axisymmetric jets, J. Fluid Mech. 14 (1962), 529-551.
5. Binder, G. and M. Favre-Marinet, Mixing improvement in pulsating turbulent jets, in Fluid Mechanics of Mixing (E. M. Uram and V. W. Goldschmidt, eds.), Am. Soc. Mech. Engrs., New York, 1973, 167-172.
6. Bouchard, G. E. and Reynolds, W. C., Control of vortex pairing in a round jet, Bull. Am. Phys. Soc. 23 (1978), 1013.
7. Browand, F. K., An experimental investigation of the instability of an incompressible separated shear layer, J. Fluid Mech. 26 (1966), 281-307.
8. Brown, G. L. and A. Roshko, On density effects and large structure in turbulent mixing layers. J. Fluid Mech. 64 (1974), 775-816.
9. Chan, Y. Y. and J. T. T. Templin, Supression of spatial waves by distortion of the jet velocity profile, Phys. Fluids 17 (1974), 2124-2125.
10. Chandrsuda, C., R. D. Mehta, A. D. Weir and P. Bradshaw, Effect of free-stream turbulence on the large structure in turbulent mixing layers, J. Fluid Mech. 85 (1978), 693-704.

11. Corrsin, S., Investigations of flow in an axially symmetric heated jet of air, Natn. Advis. Comm. Aeronaut. Adv. Conf. Rep. No. 3123, 1943 (also W-94).
12. Crow, S. C. and F. H. Champagne, Orderly structure in jet turbulence, J. Fluid Mech. 48 (1971), 547-591.
13. Durgin, W. W. and S. K. F. Karlsson, On the phenomenon of vortex street breakdown, J. Fluid Mech. 48 (1971), 507-527.
14. Elswick, Jr., R. C., Wave-induced Reynolds stress in turbulent shear layer instability. Ph.D. dissertation, The Pennsylvania State University, 1971.
15. Favre-Marinet, M. and G. Binder, Structur des jets pulsants, J. Méc. 18 (1979), 356-394.
16. Freymuth, P., On transition in a separated boundary layer, J. Fluid Mech. 25 (1966), 683-704.
17. Gaster, M., A note on the relation between temporally-increasing and spatially-increasing disturbances in hydrodynamic stability, J. Fluid Mech. 14 (1962), 222-224.
18. Gaster, M., A theoretical model of a wave packet in the boundary layer on a flat plate, Proc. R. Soc. Lond. A 347 (1975), 271-289.
19. Gatski, T. B. and J. T. C. Liu, On the interactions between large-scale structure and fine-grained turbulence in a free shear flow. III. A numerical solution, Phil. Trans. R. Soc. Lond. A 293 (1980), 473-509.
20. Ho, C. M. and L. Huang, Subharmonics and vortex merging in an unsteady shear layer, Bull. Am. Phys. Soc. 23 (1978), 1007.
21. Hussain, A. K. M. F. and W. C. Reynolds, The mechanics of an organized wave in turbulent shear flow, J. Fluid Mech. 41 (1970), 241-258.
22. Hussain, A. K. M. F. and M. F. Zedan, Effects of initial condition on the axisymmetric free shear layer: effect of the initial fluctuation level, Phys. Fluids 21 (1978), 1475-1481.
23. Hussain, A. K. M. F. and M. F. Zedan, Effects of initial condition on the axisymmetric free shear layer: effect of the initial momentum thickness, Phys. Fluids 21 (1978), 1100-1112.

24. Kelly, R. E., On the stability of an inviscid shear layer which is periodic in space and time, J. Fluid Mech. 27 (1967), 657 689.

25. Kendall, J. M., The turbulent boundary layer over a wall with progressive surface waves, J. Fluid Mech. 41 (1970), 259-281.

26. Knight, D. D., Numerical investigation of large scale structures in the turbulent mixing layer, in Proc. 6th Biennial Symposium on Turbulence, University of Missouri, Rolla, 1979.

27. Ko, D. R. S., T. Kubota and L. Lees, Finite disturbance effect in the stability of a laminar incompressible wake behind a flat plate, J. Fluid Mech. 40 (1970), 315-341.

28. Konrad, J. H., An experimental investigation of mixing in two-dimensional turbulent shear flows with applications to diffusion limited chemical reactions. Project SQUID Tech. Rep. CIT-8-PU, California Institute of Technology, 1976.

29. Launder, B. E., G. J. Reece and W. Rodi, Progress in the development of a Reynolds-stress turbulence closure, J. Fluid Mech. 68 (1975), 537-566.

30. Liepmann, H. W., Aspects of the turbulence problem, Second Part, Z. A. M. P. 3 (1952), 407-426.

31. Liepmann, H. W., Free turbulent flows, in Mécanique de la Turbulence (Coll. Intern. du CNRS à Marseille), Éd. CNRS, Paris, 1962, 211-226.

32. Liu, J. T. C., Nonlinear development of an instability wave in a turbulent wake, Phys. Fluids 14, (1971), 2251-2257.

33. Liu, J. T. C., Developing large-scale wavelike eddies and the near jet noise field, J. Fluid Mech. 62 (1974), 437-464.

34. Liu, J. T. C. and A. Alper, On the large-scale structure in turbulent free shear flows, in Proc. Symp. Turbulent Shear Flows, The Pennsylvania State University, 1977, 11.1-11.11.

35. Liu, J. T. C. and P. M. Gururaj, Finite-amplitude instability of the compressible laminar wake: comparison with experiments, Phys. Fluids 17 (1974), 532-543.

36. Liu, J. T. C. and L. Lees, Finite amplitude instability of the compressible laminar wake. Strongly amplified disturbances, Phys. Fluids 13(1970), 2932-2938.
37. Liu, J. T. C. and L. Merkine, On the interactions between large-scale structure and fine-grained turbulence in a free shear flow. I. The development of temporal interactions in the mean, Proc. R. Soc. Lond. A 352, (1976), 213-247.
38. MacPhail, D. C., Turbulence in a distorted passage and between rotating cylinders. Ph.D. dissertation, University of Cambridge, 1941. (Also in Proc. 6th Int. Congr. Appl. Mech., Paris, 1946.)
39. Malkus, W. V. R., Outline of a theory of turbulent shear flow, J. Fluid Mech. 1 (1956), 521-539.
40. Mankbadi, R. and J. T. C. Liu, A study of the interactions between large-scale coherent structures and fine-grained turbulence in a round jet, Phil. Trans. R. Soc. Lond. A 298 (1980), 541-602.
41. Miksad, R. W., Experiments on the nonlinear stages of free shear-layer transition, J. Fluid Mech. 56 (1972), 695-719.
42. Miksad, R. W., Experiments in nonlinear interactions in the transition of a free shear layer, J. Fluid Mech. 59 (1973), 1-21.
43. Moore, C. J., The role of shear-layer instability waves in jet exhaust noise, J. Fluid Mech. 80 (1977), 321-367.
44. Pai, S. I., Turbulent flow between rotating cylinders. Ph.D. dissertation, California Institute of Technology, 1939. (Also, as Natn. Advis. Comm. Aeronaut. Tech. Note no. 892, 1943.)
45. Papailiou, D. D. and P. S. Lykoudis, Turbulent vortex streets and the entrainment mechanism of the wake, J. Fluid Mech. 62 (1974), 11-31.
46. Patnaik, P. C., F. S. Sherman and G. M. Corcos, A numerical simulation of Kelvin-Helmholtz waves of finite amplitude, J. Fluid Mech. 73 (1976), 215-240.
47. Reynolds, O., On the dynamical theory of incompressible viscous fluids and the determination of the criterion, Phil. Trans. R. Soc. Lond. A 186 (1895), 123-164.

48. Reynolds, W. C., Large-scale instabilities of turbulent wakes, J. Fluid Mech. 54 (1972), 481-488.
49. Roshko, A., On the development of turbulent wakes from vortex streets, Ph.D. dissertation, California Institute of Technology, 1952. (Also, as Natn. Advis. Comm. Aeronaut. Rep. no. 1191, 1954.)
50. Roshko, A., Structure of turbulent shear flows: A new look, A. I. A. A. J. 14 (1976), 1349-1357.
51. Sato, H., Further investigation on the transition of two-dimensional separated layer at subsonic speeds, J. Phy. Soc. Japan 14 (1959), 1797-1810.
52. Sato, H. and K. Kuriki, The mechanism of transition in the wake of a thin flat plate placed parallel to a uniform flow, J. Fluid Mech. 11 (1961), 321-352.
53. Schubauer, G. B. and H. K. Skramstad, Laminar boundary layer oscillations and transition on a flat plate, Natn. Advis. Comm. Aeronaut. Rep. no. 909, 1948.
54. Stuart, J. T., On the nonlinear mechanics of hydrodynamic stability, J. Fluid Mech. 4 (1958), 1-21.
55. Stuart, J. T., Nonlinear effects in hydrodynamic stability, in Proc. Xth Int. Congr. Appl. Mech., Stresa, 1960, Elsevier, Amsterdam, 1962.
56. Taneda, S., Downstream development of wakes behind cylinders, J. Phys. Soc. Japan 14 (1959), 843-848.
57. Taylor, G. I., Stability of a viscous liquid contained between two rotating cylinders, Phil. Trans. R. Soc. Lond. A 223 (1923), 289-343.
58. Thomas, A. S. W. and G. L. Brown, Large structure in a turbulent boundary layer, in Proc. 6th Australasian Hydraulics and Fluid Mechanics Conference, Adelaide, Australia, 5-9 December 1977, 407-410.
59. Thorpe, S. A., Experiments on the instability of stratified shear flows-miscible fluids, J. Fluid Mech. 46 (1971), 299-319.
60. Townsend, A. A., The Structure of Turbulent Shear Flow, Cambridge University Press, 1956.

61. Townsend, A. A., The Structure of Turbulent Shear Flow, Cambridge University Press, 1956.

62. Whitham, G. B., Linear and Nonlinear Waves, John Wiley and Sons, New York, 1974.

63. Winant, C. D. and F. K. Browand, Vortex pairing: the mechanism of turbulent mixing-layer growth at moderate Reynolds number, J. Fluid Mech. 63 (1974), 237-255.

The most stimulating hospitality of the Department of Mathematics, Imperial College, London, in particular, that of J. T. Stuart, during my sabbatical leave in 1979/80 is gratefully acknowledged. This work was partially supported by the United Kingdom Science Research Council through its Senior Visiting Fellow Programme, the Fluid Mechanics Program of the National Science Foundation through Grant CME-78-22127-01 and the Fluid Dynamics Program of the Office of Naval Research. I am most grateful to Sandra Spinacci for the preparation of the photo-ready manuscript.

The Division of Engineering
Brown University
Providence, Rhode Island 02912
U.S.A.

Coherent Structures in Turbulence

J. L. Lumley

1. INTRODUCTION.

During the past decade there has been a very active interest in coherent structures. An excellent reference for this work is the paper of Cantwell (1981).

The general awareness of the existence of organized structures in turbulent motions can probably be traced to the work of Townsend (1956), who referred to a " double structure" of turbulent flows, consisting of more or less organized "big eddies", and a less well-organized, smaller scale, background turbulence. Townsend's work referred to so-called fully developed turbulence; that is, in general, flows that had attained selfsimilarity and were consequently many characteristic dimensions (body diameters, etc) from their origin. Townsend found that the energy in the big eddies was rather small, say of the order of one-fifth of the total turbulent energy.

The coherent structures that have attracted the interest of the turbulence community recently differ somewhat from Townsend's big eddies. For the most part, the flows that have been examined recently are younger, closer to their origins, and it has been found (usually by the use of flow visualization) that there are often present in these flows structures much more organized and more energetic than those present in fully developed turbulence. This is a matter of some controversy at the moment: one school (Brown & Roshko, 1971, 1974;

ISBN 0-12-493240-1

Roshko, 1981; Cantwell, 1981; Browand & Troutt, 1980) appears to feel that these structures are more characteristic of all turbulence than we previously thought; that we overlooked the presence of such structures in turbulence previously examined principally because we did not use visualization, and because we made measurements using ordinary statistical approaches, as opposed to the conditioned sampling widely used by this group. The claim (to which there is some truth) is, that the usual statistical approaches tend to smear out, and thus conceal, the organization truly present in the flow, which can be identified only by the use of conditioned sampling and visualization.

The opposing school, (for example: Bradshaw, 1966; Hussain, 1981) feels that there is substantial evidence that the degree of organization in these flows decreases as the flows age ;that one reason we were not previously aware of the existence of these well organized, energetic structures is that we never measured in the early part of turbulent flows, saying that the flows were not yet fully developed; that, in fact, the well-organized structures observed may be attributable in part to the care with which these flows have been set up - that is, if extreme care is taken to remove adventitious disturbances from the oncoming flow (as is now more usual than formerly), the instability of the initial flow will be of a single type, and the transition flow will be dominated by the nonlinear evolution of this instability. The initial development of the turbulence will probably be influenced for some downstream distance by these structures.

Whether these structures are the same as those that will be present when the flow is fully developed is difficult to say; certainly there is evidence that in some cases they are not (compare Bevilaqua & Lykoudis, 1971 and Payne & Lumley, 1967). There is some feeling that the structures present in the early flow are probably more characteristic of the initial instability, while those present in the fully developed turbulence are probably characteristic in some sense of the fully developed flow (represent, say, some type of instability of that flow), and are consequently not necessarily the same, unless the initial flow has been set up so that the dominant instability present there is of the same type as that present in the fully developed flow. There is evidence for this in the development of the flat plate wake of Chevray & Kovasznay (1969), which is formed without vortex shedding, and is quite different from the cylinder wake of Townsend (1947); the same is true of the spheroid wake of Chevray (1968) and the porous disk of Bevilaqua & Lykoudis (1971), which do not contain vortical structures, in contrast to the sphere wake of Bevilaqua & Lykoudis (1971). This is also consistent with the results of second order modeling (which does not include the effect of large coherent structures), which forms a progression with the experimental results cited: from the wake with structures, to that without, to the model, the latter two being quite close together (see Taulbee & Lumley, 1981). This suggests that the wake supposedly without structures actually has weak ones.

Roshko (1981) points out that the mixing layer (perhaps uniquely) will appear thin to disturbances of sufficiently large scale at any stage of its development, and consequently should be subject to the same type of (Kelvin-Helmholtz) instability at all ages; he would thus expect to find energetic, well-organized structures of the same type at all ages. The same might be said of other nearly-parallel shear flows; the wake, for example, should be even thinner than the shear layer. The instability here, however, would not be of the Kelvin-Helmholtz type, but would presumably be associated with the inflectionary profile; the inflection points are buried in the small-scale turbulence, however, and this instability is probably strongly damped by the momentum transport of this fine-grained turbulence. This might be a reason why energetic, well-organized structures would not be observed in the late, equilibrium (similarity) wake. The mixing layer may be unique in this respect, although one might expect a certain amount of disorganization to creep in with age - that is, the older mixing layer is already quite disturbed by turbulence of all scales, including non-linearly evolved older three-dimensional somewhat disorganized structures; although subject to a Kelvin-Helmholtz type of instability, this could be damped and modified by the momentum transport of the other eddies present, resulting in a somewhat weaker and less-well organized coherent structure than one might find in an early mixing layer which has just undergone transition. The measurements of Browand & Troutt (1980) appear to support this view; they find that the mixing layer begins to evolve in a self-similar way after some time, and that although there are structures present that extend across the flow, they are somewhat disorganized (skewed and branched). It is certainly true that ordinary statistical measures do not make clear the cross-stream extent of these structures.

In naturally occuring flows, and flows in machines, the possibility of encountering fully developed turbulence is relatively remote (with the exception of the boundary layer). Jet noise is produced by the part of the jet within a few diameters of the nozzle; mixing layers of interest in the design of slots and flaps are quite close to their origin. We may consequently expect that these flows will be characterized by much more energetic and well-organized structures than are present in the academically interesting fully-developed turbulence. Whether these structures will be as well-organized as those currently under investigation by the coherent-structures community is an open question; it seems likely that the flows occuring in machines and in nature will be much more disturbed than those currently under investigation, and hence that the structures will lose some of their organization.

Whether more or less organized, and more or less energetic, it is clear that some form of organized structure is present in turbulent shear flows of all types. While it is of academic interest to calculate these structures in fully developed turbulence, it is not usually of practical importance; in most cases, the organized structures contain a sufficiently small part of the total energy, and play a sufficiently secondary role in transport, that calculations made, say, by second order modeling, which ignores the existence of these structures, are satisfactory. In the flows of more practical importance, however, which are for the most part young, the organized structures play an undeniable role, and a calculation technique that takes their presence into account is necessary. We will discuss such a technique below.

We have previously proposed a technique for identifying these structures (Lumley, 1967, 1970; Payne, 1966; Bakewell, 1966) which uses conventional statistical approaches. I am a little uneasy about the use of conditioned sampling, since it is necessary to introduce the prejudices of the experimenter in order to supply the condition. To quote Loren Eisely (1979; p. 199), "Man, irrespective of whether he is a theologian or a scientist, has a strong tendency to see what he hopes to see". I will show below that one can find in statistical data irrelevant structures with high probability; I cannot call them non-existent, since they are there, but they are formed by chance juxtaposition of other, relevant, structures, and have no significance. They are rather like the birds in an Escher woodcut (Hofstadter, 1980), that are formed by the spaces between other birds. Of course, I cannot generalize; in the hands of a superb experimenter, probably even the most unsatisfactory technique will not lead one astray. In addition, the experimentalist cannot be a _tabula rasa_, as Liepmann points out (private communication): if the experiment is to be a success, the experimentalist must introduce his prejudices in one way or another. There are, finally, experimental situations that are clearer physically than others, where there is less likeihood of error. Other things being equal, however, I prefer a less prejudicial technique for identifying the presence of coherent structures; specifically, I prefer the technique I will describe below.

The use of conventional statistical approaches in general has recently been very much criticised. Laufer, for example, has said (during the present meeting), "[The use of conventional statistical approaches] never got us anywhere". I find this an immoderate view, and feel that its implication, that these approaches should therefore be abandoned, is unwarrented. It is probably true that conventional statistical approaches did not help much to elucidate the organized structure of shear flows, but that is not the only problem in turbulence; I believe they have been very helpful in other respects. That they have not been helpful in this respect I believe is attributable not to an inherent shortcoming in these approaches, but to their misuse. Relatively few measurements of moments higher than second have been made, and these have not usually been tied to the physics. We shall show below that much of the information which is felt to be missing from the conventional statistical descriptions may be found in the third and fourth moments. Cantwell (1981) says, "To an investigator of the 1920s or 1930s, turbulence was essentially a stochastic phenomenon...", and later, "The last twenty years ...have seen a growing realization that ...most turbulent shear flows are dominated by ...motions that are not random". Cantwell to the contrary notwithstanding, most turbulence investigators still consider turbulence to be essentially stochastic, or random, in nature; this does not mean that there are not organized structures, but that these structures occur at times and places, and with strengths and shapes, that jitter, the extent of the jitter depending on the flow situation. It is this jitter that requires the use of what is called boot-strapping by the coherent structures people - the search for a delay that will maximize the correlation. This technique works as well, of course, to improve the correlation of an irrelevant signal as of a relevant one, and has a history in statistics of being extremely dangerous (in the sense that, if a finite piece of the variable is in question, a lag can be found at which the correlation is as good as desired, regardless of the relation or lack of it).

Cantwell (1981) says further that "...there does not exist a unique relationship between the correlation tensor and the unsteady flow that produces it". While this is formally true, it is misleading. As we shall show below, the correlation tensor gives a decomposition of the unsteady flow that produces it which is unique except for phase, which must be determined from third moments. That is, the structure of the organized motions (except for phase) is determined by the correlation tensor - other aspects of the occurrence of the motions must be determined from higher order statistics.

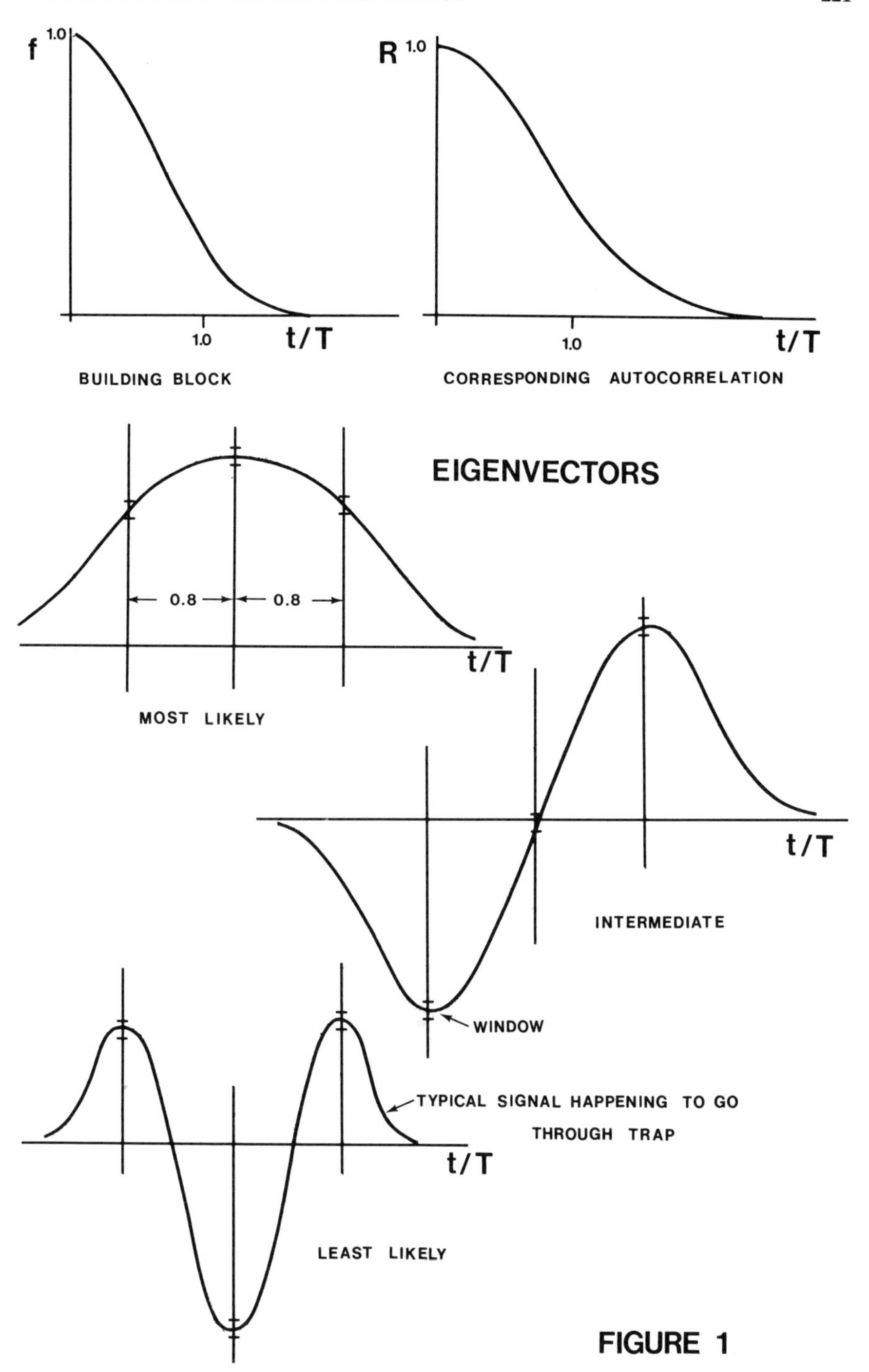
f
1.0
1.0
t/T
BUILDING BLOCK
R
1.0
1.0
t/T
CORRESPONDING AUTOCORRELATION
EIGENVECTORS
0.8
0.8
t/T
MOST LIKELY
t/T
INTERMEDIATE
WINDOW
TYPICAL SIGNAL HAPPENING TO GO THROUGH TRAP
t/T
LEAST LIKELY

FIGURE 1

2. AN OBJECT LESSON.

Let us examine the possibility of "identifying" an irrelevant structure in a stochastic signal. We may consider a signal which is composed of building blocks in the following manner:

$$u(t) = \int f(t-\tau)g(\tau)d\tau \tag{2.1}$$

where f(t) is a deterministic function and g(t) is a stochastic function, independent in non-overlapping intervals and with a Gaussian probability density for amplitude. For our function f(t) we may select a Gaussian bell-shaped curve (figure 1), so that a realization of the random function u(t) might look like figure 2:

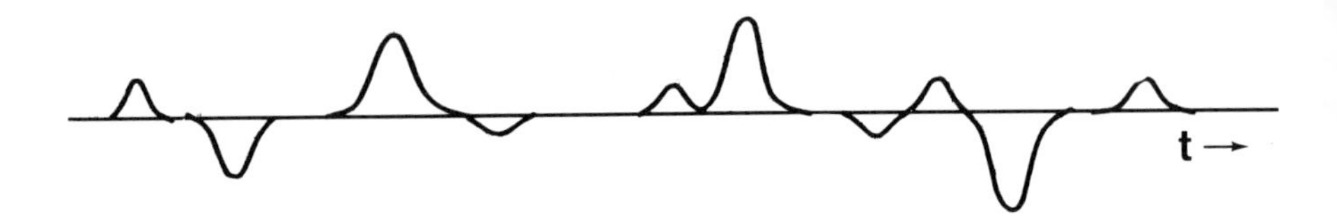

Figure 2. A realization of a function composed of randomly located bell-curves with Gaussian amplitudes.

To approach this problem, we may now consider the probability that a realization of this function passes through a series of windows spaced equally along the axis. Call the times at which the windows are located t_k, $k = 1,\ldots,n$, and the values of the function at those times $u(t_k) = u_k$. What we wish is the probability that u_1 has approximately one value, u_2 has another, and so on, or

$$\Pr\{c_k < u_k < c_k+\Delta c_k,\ k = 1,\ldots,n\} \sim \exp[-R^{-1}_{ij}c_i c_j/2] = P\{\underline{c}\} \tag{2.2}$$

where the form may be written down directly since we know that the amplitude behavior of u(t) is Gaussian. The quantity appearing in the exponent of equation (2.2) is the autocorrelation matrix:

$$R_{ij} = \overline{u_i u_j} \tag{2.3}$$

We may now ask what trap, or collection of windows, is most probable. We would expect that this would be something like the building blocks of which we constructed the function. If we extremize the probability for a fixed mean-squarevalue of the trap $c_i c_i = c^2$ we obtain a classical eigenvalue equation for the autocorrelation matrix:

$$R_{ij} X_j^{(k)} = \lambda^{(k)} X_i^{(k)}, \qquad X_i^{(k)} X_i^{(\ell)} = \delta_{k\ell} \tag{2.4}$$

For our particular case, with traps consisting of three windows spaced 0.8 integral scales apart, we have computed the eigenvectors, which are shown in figure 1. It is seen that the most likely is indeed similar to the building blocks of u(t). The second and third eigenvectors can be seen to correspond to the accidental occurrence of either two or three of the fundamental building blocks in juxtaposition, with alternating signs. Any trap can be constructed of these building blocks. In particular, a trap that is proportional to an eigenvector has an especially simple probability:

$$P\{c\underline{X}^{(k)}\} = \exp[-c^2/2\lambda^{(k)}] \tag{2.5}$$

Finally, we may integrate over the amplitude, to obtain the probability that u(t) passes through a trap of given shape regardless of amplitude:

$$\int P\{c\underline{X}^{(k)}\}dc \sim [\lambda^{(k)}]^{1/2} \tag{2.6}$$

The eigenvalues of our three-window trap are given by

$$\lambda^{(1)} = 1.92, \lambda^{(2)} = 0.870, \lambda^{(3)} = 0.214 \tag{2.7}$$

Hence, the ratio of the probability of the most likely to that of the next most likely is only 1.49, and to the least likely is only 3.00. If the experimenter set his trap to catch a function like the third eigenvector, which is the least likely event he could look for, he will find one for every three of the (most likely) building blocks.

As the number of windows in the trap is increased, the calculation, of course, becomes more complex. The result, however (the square root of the ratio of the first to the second eigenvalues), does not seem to be particularly sensitive to the number of windows in the trap, so long as the same range of the variable is covered, nor does it seem to be particularly sensitive to the exact form of the autocorrelation function (i.e.- to the exact form of the building block).

The moral I would like to draw from this is that it is dangerous to go hunting for a rare bird in a random field, since the field is filled with not-so-rare birds which only appear to exist.

3. A NON-PREJUDICIAL APPROACH

Let us consider a vector function of a single variable, $u_i(x_2)$, and let us suppose that $\underline{u}$ is inhomogeneous, and that the energy is integrable:

$$\int \overline{u_i u_i} dx_2 < \infty \tag{3.1}$$

We will deal with the homogeneous situation, and with combined situations, later. This material and that in sections 4. and 7. can be found in Lumley (1967, 1970). Now, when we think we see an organized structure in the random function $\underline{u}$, one way of testing this hypothesis is to take the correlation of the random function with the proposed candidate structure; find out, that is, how nearly parallel the proposed structure and the random function are in function space:

$$a = \int \phi_i^* u_i dx_2 / [\int \phi_i \phi_i^* dx_2]^{1/2} \tag{3.2}$$

where ( )* indicates the complex conjugate, which we include for generality. The quantity in (3.2) has been normalized by the amplitude of the candidate structure so that the projection of the candidate on the random function will not be affected by the amplitude of the candidate, only by its shape. If we adopt a probabilistic approach, we are interested not in the value of (3.2) in a single realization, but in some statistical measure of (3.2) over the ensemble of realizations. The simplest statistical measure is the mean square of the absolute value. Now, instead of guessing at various candidate structures and testing each one, we can ask if there is a structure that will maximize the mean square magnitude of the projection. This is a well-defined extremization problem:

$$\int R_{ij}(x_2,x_2')\phi_j^{(n)}(x_2')dx_2' = \lambda^{(n)}\phi_i^{(n)}(x_2) \tag{3.3}$$

where the kernel is the autocorrelation matrix

$$R_{ij}(x_2,x_2') = \overline{u_i(x_2)u_j(x_2')} \tag{3.4}$$

There is not just one, but a denumerable infinity of solutions which are orthogonal, and can be normalized:

$$\int \phi_i^{(p)}\phi_i^{*(q)}dx_2 = \delta_{pq} \tag{3.5}$$

which is to say that the structures of various orders have nothing in common with each other. The random function $\underline{u}$ can be represented in terms of the eigenfunctions:

$$u_i(x_2) = \sum a_n\phi_i^{(n)}(x_2), \quad a_n = \int u_i\phi_i^{*(n)}dx_2 \tag{3.6}$$

where the random coefficients of different orders are uncorrelated, and their mean square value is given by the eigenvalues:

$$\overline{a_n a_m^*} = 0,\ n \neq m; \ = \lambda^{(n)},\ n = m \tag{3.7}$$

This is probably the most significant part of the representation theorem: the random function may truly be reconstructed from these structures with random coefficients. Note that the representation converges optimally fast - the first coefficient is (in mean square) as large as possible; of the remainder that could not be incorporated in the first term (because it was orthogonal to it) the coefficient of the next term is as large as possible, and so forth. There are a number of other properties that need not concern us here (which may be found in Lumley 1970): there is a representation for the autocorrelation matrix in terms of the eigenfunctions, the eigenvalues are positive (from their definition), and their sum converges, there are straightforward ways of calculating the eigenvalues and eigenfunctions, and even of estimating how many terms in the series are necessary for a representation (essentially the ratio of the length scale characteristic of the inhomogeneity to that of the energy containing eddies, a ratio which is seldom more than three).

This is a non-prejudicial way of extracting organized structures from an inhomogeneous random function, on an energy-weighted basis; that is, the structures are those that contribute most to the energy. This representation is known in the literature of probability theory as the proper orthogonal decomposition theorem.

4. HOMOGENEOUS DIRECTIONS

The representation of section 3, while useful, does not directly address the situation we discussed in section 2, in which the random function was stationary or homogeneous. If we apply the decomposition of section 3 to such a situation, we find that the eigenfunctions are no longer discrete, but there is now a continuous spectrum of them, and they are in fact the Fourier modes. The proper orthogonal decomposition theorem reduces to the harmonic orthogonal decomposition theorem. While a Fourier representation is, of course, useful for many purposes, it suffers from the disadvantage that the eigenfunctions are not confined principally to one region of space or time; they are not eddies, in the sense in which one usually thinks of them, since they do not correspond to physical entities that one can see in flow visualization.

Fortunately, another decomposition exists which is appropriate for homogeneous directions. This is an outgrowth of the shot-effect expansion (see Rice, 1944; Lumley, 1970). Any homogeneous function may be written as

$$u_i(x) = \int f_i(x-x')g(x')dx' \tag{4.1}$$

where $f_i(x-x')$ is a deterministic function and $g(x')$ is a stochastic function. One has a certain amount of freedom in choosing how much of the behavior of $\underline{u}$ to put in $\underline{f}$, and how much to put in g. The usual choice, and the one that seems most convenient, is to make g white; that is, uncorrelated in non-overlapping intervals. Specifically, if we pick

$$\overline{g(x)g(x')} = \delta(x'-x) \tag{4.2}$$

then we have Campbell's theorem:

$$R_{ij}(\xi) = \int f_i(x)f_j(x+\xi)dx \tag{4.3}$$

If we take Fourier transforms, we can write for the spectrum

$$\Phi_{ij}(k) = \hat{f}_i\hat{f}_j^* \tag{4.4}$$

where we are indicating the Fourier transform of $\underline{f}$ by a circumflex. It is clear that (4.4) determines $\underline{f}$ or its Fourier transform to within a phase angle:

$$\hat{f}_\alpha = \Phi_{\alpha\alpha}^{1/2}\exp[i\theta(k)] \tag{4.5}$$

This is not to say that the phase angle is undeterminable, simply that it is undeterminable from second order statistics. In the next section we will show how the phase angle may be determined. Once this has been done, we will have exactly what we want - a representation of the random function as a series of coherent structures occurring at stochastic locations with stochastic strengths. Just as we cannot determine the phase from the second order statistics, we also cannot say anything about the way in which the structures repeat - with overlapping, or without overlapping, and with what periodicity, etc., on the basis of second order statistics. Again, this is not to say that this information is undeterminable, simply that it cannot be determined from second order statistics. In section 5 we will consider the retrieval of phase information, and in section 6 the retrieval of information on overlap and spacing.

5. RETRIEVING PHASE INFORMATION

Let us consider a signal composed of building blocks as in (2.1). We will make a specific choice of f(t) as

$$f(t) = (d/dt)[2\pi]^{-1/2}\sigma^{-1}\exp[-t^2/2\sigma^2] \quad (5.1)$$

so that the building blocks are odd functions; these might correspond, for example, to measurement of transverse velocity due to a random sprinkling of vortices. We will take the function g(t) to be independent in nonoverlapping intervals (so that these building blocks are truly distributed chaotically), and to have the following probability structure:

$$g(t)dt = 1 \text{ with prob } \mu dt, = 0 \text{ with prob } 1-\mu dt \quad (5.2)$$

The vortices are all of the same sign and strength, now; although the weighting function does not have zero mean, the building-block function does, so that the function u(t) does also. Now, we obtain from Campbell's theorem

$$S = \hat{f}\hat{f}^* = \omega^2\exp[-\sigma^2\omega^2] \quad (5.3)$$

where S is the spectrum. Hence, we have for the transform of f

$$\hat{f} = S^{1/2}\exp[i\theta], \quad S^{1/2} = |\omega|\exp[-\sigma^2\omega^2/2] \quad (5.4)$$

Actually, although we cannot determine the phase angle from the spectrum, we know from the definition of f in (5.1) that it is -90° for positive frequencies and +90° for negative frequencies. That the proper phase angle makes a considerable difference can be seen from an examination of figure 3; if f is reconstituted with the proper amplitude but with zero phase angle, the resulting function is even, rather than odd, and would give quite a misleading impression.

We can recover the phase information from the triple correlation. By an extension of Campbell's theorem we can write (with our assumptions on the behavior of g(t)):

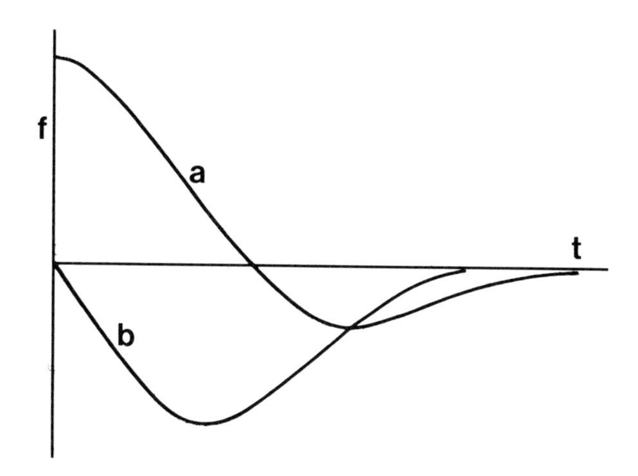

Figure 3. The function f reconstituted with the correct amplitude, but in the case (a)with zero phase angle. (b) has the correct phase.

$$\overline{u(t)u(t+\tau_1)u(t+\tau_2)} = \mu\int f(x)f(x+\tau_1)f(x+\tau_2)dx = R_2(\tau_1,\tau_2) \qquad (5.5)$$

Taking the double Fourier transform of this, we obtain the bi-spectrum described and discussed by Rosenblatt and his co-workers (Brillinger & Rosenblatt, 1967a,b; Rosenblatt, 1966; Rosenblatt & Van Ness, 1965):

$$S_2(\omega_1,\omega_2) = \int\int \exp[i\omega_1\tau_1+i\omega_2\tau_2]R_2(\tau_1,\tau_2)d\tau_1 d\tau_2 \qquad (5.6)$$

$$= \mu\hat{f}(\omega_1)\hat{f}(\omega_2)\hat{f}^*(\omega_1+\omega_2) \qquad (5.7)$$

$$= \mu S^{1/2}(\omega_1)S^{1/2}(\omega_2)S^{1/2}(\omega_1+\omega_2) \times\exp[i\{\theta(\omega_1)+\theta(\omega_2)-\theta(\omega_1+\omega_2)\}] \qquad (5.8)$$

Notice that if the phase angle is proportional to the frequency, the exponent vanishes. However, an exponent proportional to frequency corresponds to a simple time delay; the shape of the building block is not affected, but the time at which the center occurs is shifted. In the general case, we can move the three frequencies as close together as possible, and solve (5.8) to obtain the phase angle of f in terms of that of the bi-spectrum. We can make the upper and lower frequencies define essentially the edges of an eddy (see Tennekes & Lumley, 1972) if we make

$$\omega_1 = \alpha\omega, \ \omega_2 = \omega, \ \omega_1+\omega_2 = 1/\alpha, \ 1/\alpha = 1+\alpha, \quad \alpha = 0.62 \tag{5.9}$$

The upper and lower frequencies are then equally divided logarithmically from the center frequency. If we define

$$\theta(\omega) = a+b\omega+c\omega^2+d\omega^3+\dots \tag{5.10}$$

$$\psi = \theta(\alpha\omega) + \theta(\omega) - \theta(\omega/\alpha)$$

then we find

$$\psi = a-2c\alpha\omega^2-d(3\alpha^2+3\alpha)\omega^3-\dots \tag{5.11}$$

so that all the coefficients in the expansion of the phase angle can be determined except for the linear term (which is irrelevant, as discussed above). It is necessary, of course, to carry out the analysis separately for positive and negative frequencies. In our case, c and d are zero, and a takes on the value -90° for positive frequencies and +90° for negative, so that the phase angle is determined.

We have admittedly simplified things a bit by our assumptions regarding the structure of u(t) in this example. In more complex situations, the right hand side of (5.7) would be multiplied by the bi-spectrum of g(t). Since in a general representation, only the second order properties of g have been selected up to this point, we can specify g(t) more precisely by making further choices now, consistent with the measurements.

6. OVERLAP AND RECURRENCE

Hussain (1981) refers to the suggested decomposition (2.1), but says that it is, unfortunately, not useful because it implies that the eddies overlap. In fact, this is not true. The only assumption that is made is that the function $g(t)$ is uncorrelated in non-overlapping intervals. This does not imply, however, that it is independent in non-overlapping intervals. It is well-known (see, for example, Tennekes & Lumley 1972) that lack of correlation and independence are not the same thing. Let us consider the probability density for $g(t)$:

$$B(u,v;\tau)dudv = Pr\{u<g(t)dt<u+du, v<g(t+\tau)dt<v+dv\} \tag{6.1}$$

The autocorrelation of $g(t)$ is given by

$$\overline{g(t)g(t+\tau)}dt^2 = \int uvBdudv \tag{6.2}$$

If the probability density B is symmetric for non-zero lags, then $g(t)$ will be uncorrelated. It is fairly easy to display a B which suppresses the occurrence of a second structure for a time T, but is statistically independent thereafter. In the following example, by making the starred standard deviation as small as we like relative to the unstarred, we may suppress as much as we like the occurrence of two structures closer than T:

$$B(u,v;\tau) = \{\exp[-u^2/2\sigma^2 - v^2/2\sigma_*^2] + \exp[-u^2/2\sigma_*^2 - v^2/2\sigma^2]\}/4\pi\sigma\sigma_*, 0<t<T \tag{6.3}$$

$$= \{\exp[-u^2/2\sigma^2]/\sigma + \exp[-u^2/2\sigma_*^2]/\sigma_*\} \times \{\exp[-v^2/2\sigma^2]/\sigma + \exp[-v^2/2\sigma_*^2]/\sigma_*\}/8\pi, \quad t>T \tag{6.4}$$

This is sketched in figure 4 for a moderate ratio of the standard deviations.

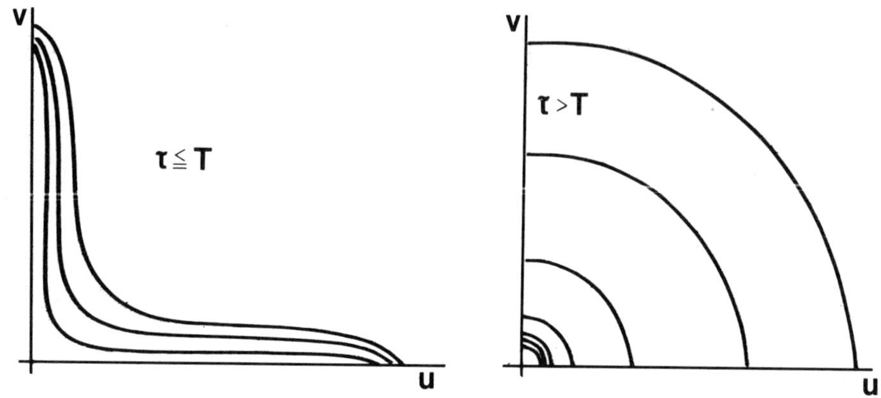

Figure 4. Qualitative sketch of iso-probability contours for the joint density of equations (6.3) and (6.4), for a moderate value of the ratio of standard deviations.

Integrating over one of the variables, we obtain the expression for the single density:

$$\int B(u,v;\tau)dv = \{\exp[-u^2/2\sigma^2]/\sigma + \exp[-u^2/2\sigma_*^2]/\sigma_*\}/2(2\pi)^{1/2} \qquad (6.5)$$

and it is evident that the function has an intermittency factor of 0.5. That is, the function g(t) is essentially on 50% of the time and off 50% of the time. In figure 5, we show a rough sketch of this density. Of course, this choice of densities by no means exhausts the possibilities; densities could be easily constructed in which the loss of suppression was gradual, or in which the density did not become statistically independent after the time T, but became increasingly stimulative, so that as time went on, a second occurrence would become more and more certain. These are just two among many interesting variations. So far as data processing is concerned, once the phase of f(t) is known, relation (2.1) can be Fourier transformed and the individual values of the transform of u(t) divided by the transform of f(t) to give the transform of g(t), which can then be Fourier transformed back to real space; now, from the real g, the probability density can be obtained.

In the case of our simple example, the information on recurrence is most easily obtained from higher moments of g(t). Specifically, the fourth moment is the lowest one to contain this information. If we form the autocorrelation of the square

$$\rho_2 = \overline{[g^2(t)-\overline{g^2(t)}][g^2(t+\tau)-\overline{g^2(t+\tau)}]}/\overline{[g^2(t)-\overline{g^2(t)}]^2} \quad (6.6)$$

we obtain (for a vanishingly small ratio of the standard deviations) the sketch in figure 6:

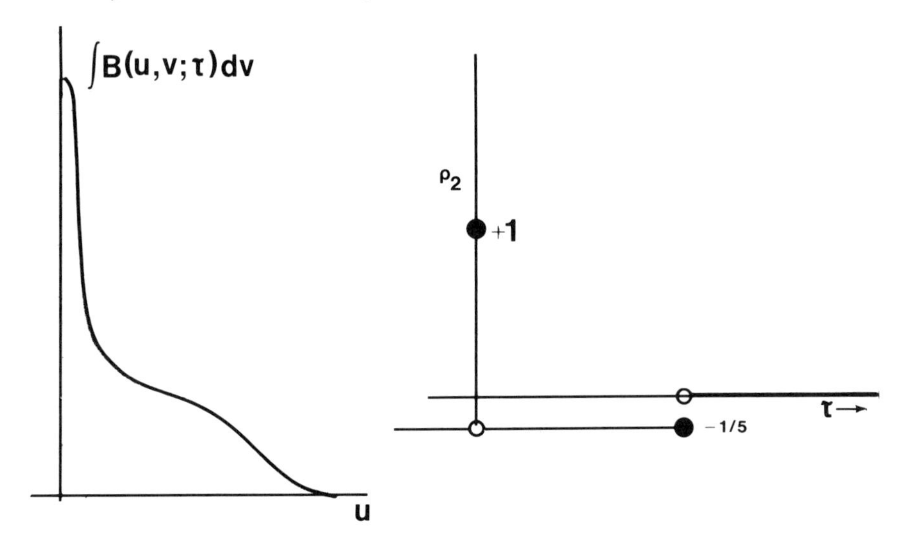

Figure 5, (Left). A sketch of the single density corresponding to equation (6.5). Figure 6, (Right). A sketch of equation (6.6) for the model of eq. (6.3-4).

The fact that this correlation drops to -1/5 instead of -1 is only a reflection of the fact that it is a fourth order correlation; the value of -1/5 indicates perfect anti-correlation, or complete suppression, if you will. If, instead of becoming statistically independent, the process became stimulative (as opposed to suppressive), we would expect the correlation to rise to plus one. The moral is, that a model such as (2.1) makes no suppositions regarding overlap and recurrence; this information is contained in the higher statistics.

7. NEARLY PARALLEL SHEAR FLOWS

In jets, wakes, shear layers, boundary layers, and so forth, we have a situation that is approximately homogeneous in the x_1 and x_3 directions, and approximately stationary, while being of integrable energy in the x_2 direction. In such a flow, we must combine the representation of section 2. with that of section 3. We can write for the fluctuating velocity field

$$u_i = \sum_n \int \exp[i(k_1x_1+k_3x_3+\omega t)]\phi_i^{(n)} a_n dk_1 dk_3 d\omega \tag{7.1}$$

where phi is the deterministic eigenfunction and a_n is the random coefficient. The coefficients of different orders are uncorrelated:

$$\overline{a_n a_m^*} = \lambda^{(n)} \delta_{nm} \tag{7.2}$$

and the mean square value is the eigenvalue:

$$\int \Phi_{ij}(x_2,x_2')\phi_j^{(n)}(x_2')dx_2' = \lambda^{(n)}\phi_i^{(n)}(x_2) \tag{7.3}$$

where the kernel is the cross-spectral density (we are suppressing the wavenumbers and frequencies):

$$\overline{\hat{u}_i(k_1,\ldots)\hat{u}_j^*(k_1',\ldots)} = \delta(k_1-k_1')\ldots\Phi_{ij} \tag{7.4}$$

$$u_i = \int \exp[i(k_1x_1+\ldots)\hat{u}_i dk_1 dk_3 d\omega$$

Again, the eigenfunctions are orthogonal, and can be normalized:

$$\int \phi_i^{(p)}\phi_i^{(q)*}dx_2 = \delta_{pq} \tag{7.5}$$

We may identify the organized structure by

$$\int \exp[i(k_1x_1+\ldots)\phi_i^{(1)} a_1 dk_1 \ldots = \int f_i(x_1-\xi_1,\ldots)g(\xi_1,\ldots)d\xi_1\ldots \tag{7.6}$$

where the weighting function is uncorrelated in non-overlapping intervals:

$$\overline{g(\xi_1,\ldots)g(\xi_1',\ldots)} = \delta(\xi_1-\xi_1')\ldots \tag{7.7}$$

Finally, the deterministic function which is sprinkled randomly is given by:

$$f_i = \int \exp[i(k_1x_1+\ldots)][\lambda^{(1)}/(2\pi)^3]^{1/2}\phi_i^{(1)}dk_1\ldots$$
$$= \int \psi_i^{(1)}dk_1\ldots \text{ say} \tag{7.8}$$

From a physical point of view, these shear flows present the most interesting case. If we are to develop equations to predict the form of these organized structures, it is for the function of (7.8) that we must do so.

8. DYNAMICAL EQUATIONS

We may approach the dynamical equations directly. If we begin with the 1-component of the equation for the fluctuating velocity:

$$\dot{u}_1+u_{1,1}U+U'u_2+u_{1,j}u_j = p_{,1}/\rho+\nu u_{1,jj} \tag{8.1}$$

Multiplying by a_k we obtain the Fourier transform:

$$\overline{a_k u_i} = \int \exp[ik_1x_1+\ldots]\phi_i^{(k)}\lambda^{(k)}dk_1\ldots \tag{8.2}$$

Hence, we must multiply equation (8.1) by a_k, average, take the inverse Fourier transform, and divide by the eigenvalue. This leads to the following:

$$ik_1(U+\omega/k_1)\phi_1^{(n)}+U'\phi_2^{(n)}$$
$$+\textstyle\sum_{p,q}\int D_j\phi_1^{(p)}(k_1',\ldots)\phi_j^{(q)}(k_1-k_1',\ldots)\overline{a_ka_pa_q}/\lambda^{(k)}dk_1'\ldots+\ldots \tag{8.3}$$

where we have written only the terms on the left hand side, since these are the ones that will cause difficulties. We are now faced with the classical closure problem of turbulence, though in a slightly unfamiliar form: we must somehow terminate this coupled hierarchy of equations. One possibility that comes to mind is a sort of Heisenberg approach (Monin & Yaglom, 1975), in which the higher order eigenfunctions are supposed to act like a viscosity on the lower order ones. This has greater likelihood of success here than it does in the homogeneous case (where it nevertheless gives results that are not qualitatively bad) because here there is between eigenfunctions a sort of spectral gap - successive eigenfunctions have relatively little in common with each other, being orthogonal.

There is, however, a simpler approach which may produce adequate results. We may begin from the equation for the Reynolds stress tensor, and introduce there one of the closures that is used in second order modeling, obtaining from this closed equation for the Reynolds stress an equation for the eigenfunction (from the integral equation for the eigenfunction: by multiplying by the eigenfunction and integrating). The Reynolds stress is given by

$$R_{ij} = \overline{u_i(\underline{x}',t')u_j(\underline{x},t)} \tag{8.4}$$

and the equation for the Reynolds stress (indicating only the terms for which closure is necessary; the pressure-strain correlation and the viscous terms do not require modeling in this approach):

$$\dot{R}_{ij}+\ldots+(\overline{u_i'u_ju_k})_{,k}+\ldots \tag{8.5}$$

Now, the most elementary sort of mixing length approach would replace the turbulent transport term by

$$\overline{u_i'u_ju_k} = -(R_{ij,\ell}\overline{u_k u_\ell}T)_{,k} \tag{8.6}$$

This is a slight extension of the closures used in second order modeling, since those closures do not ordinarily pertain to the Reynolds stress at two different points and times. If the implications of (8.6) are examined carefully, it is found that it is not disastrously unreasonable, except for the behavior of the time correlation, which does not decay under the assumption (8.6). We may introduce a further simplification if we note that in a first approximation the transport coefficient in (8.6) is constant over a cross-section of all the nearly parallel shear flows. Using assumption (8.6), and proceeding as described above, we may obtain an equation for the function defined by equation (7.8):

$$\dot{\psi}_i^{(1)}+\psi_{i,j}^{(1)}U_j+U_{i,j}\psi_j^{(1)} = -p_{,i}^{(1)}/\rho+(\psi_{i,\ell}^{(1)}\overline{u_k u_\ell}T)_{,k}\ , \qquad \psi_{i,i}^{(1)} = 0 \tag{8.7}$$

The viscous terms have been neglected, since they are small relative to the transport terms arising from the triple correlations. Note that this equation has the form of the linearized Navier-Stokes equation. It is not truly linear, since the transport coefficient in fact is given by

$$\overline{u_i u_j} = \sum_n \int \lambda^{(n)} \phi_i^{(n)} \phi_j^{(n)*} dk_1 \ldots \tag{8.8}$$

The equation is consequently cubic. Note that the exact equation (8.3) is quadratic; our closure scheme has changed a quadratic equation to a cubic one, and that is bound to have interesting consequences. The time scale which appears in (8.6) and in (8.7) is proportional to the ratio of turbulent energy to dissipation rate.

The attractive feature of equation (8.7) is that it may be solved by a sort of self-consistent field approximation. We take, as a first approximation, an isotropic turbulent field; the transport coefficient is then an isotropic eddy viscosity, which we take constant as a first approximation. Equation (8.7) then leads directly to the classical Orr-Sommerfeld equation; that is, we must solve the linear stability problem for the flow in question. We can accept only neutral disturbances, since the frequency and wavenumber must be real; the frequency and wavenumber arise not by an arbitrary Fourier expansion, but from the representation theorem which embodies the fact that the turbulence is stationary and homogeneous. We do not know the value of the eddy viscosity, but we know that the Reynolds number based on it must be at least above the critical value in order to have a solution with real frequency and wavenumber. If the Reynolds number is above the critical value, we will have a spread of such frequencies and wavenumbers. The disturbance will presumably grow slowly (so as not to violate the stationarity assumption); as it grows, the eddy viscosity will grow, and the value of the Reynolds number will be reduced. The most energetic possible disturbance will then be that which has reduced the Reynolds number to its minimum critical value, at which point only a single disturbance (a single wavenumber and frequency) will remain. This is the marginal stability idea that has been put forward speculatively by many writers (Malkus, 1956; Malkus & Veronis, 1958; Lessen, 1978): that the eddy viscosity should have such a value that the turbulent flow should be marginally stable to small disturbances. However, we have had to make no assumptions to arrive here, other than the closure assumption.

To implement the self-consistent field approximation, we should now construct a transport coefficient from our newly obtained critical disturbance, and solve the resulting equation again to obtain a better approximation to the critical disturbance. However, there is probably no point in this until a number of questions have been satisfactority resolved. For example: at the critical state, we are reduced to a single disturbance. This is not a true representation of reality - real flows are disturbed by a spread of wavenumbers and frequencies. How can this be arranged? We need a better closure of the Reynolds stress equation, one that better represents the time behavior of the correlation, since this surely will have an impact on the behavior of the eigenfunctions. We have not said anything about how to specify a length scale (essentially, we have a velocity scale-to specify a time scale, we need a length scale). It is not clear how to weight the eigenfunctions of various orders in reconstructing the transport coefficient - if only the first eigenfunction were present, our technique will permit us to determine the first eigenvalue, but if more than one is present, it is not clear how to determine the higher ones. Finally, it is not clear how to deal with flows, like the mixing layer, that have no minimum critical Reynolds number. Presumably many of these questions would be answered by a better closure, which would behave more sensibly. What we have here is a very crude model which nevertheless shows us a crude mechanism.

9. APPLICATIONS TO MEASUREMENTS

The decomposition of section 7. is being applied to the early part of a circular jet by W. K. George, Jr. and his colleagues at SUNY Buffalo. They hope to shed light on the nature of the coherent structures that are responsible for noise generation in this flow. This work is being carried out with the support of the U. S. Air Force Office of Scientific Research.

This same decomposition will also be applied to measurements taken in the sub- and buffer layer of a turbulent boundary layer by Siegfried Herzog and the present author. Briefly, two components of velocity have been measured at 882 point pairs. The spacing increases by factors of two in each direction, the minimum spacing in the cross-stream directions being 0.7 (in dimensionless sublayer units), while that in the streamwise direction is 9.0. Three minutes of data were taken at each location; this should be sufficient to obtain statistics with better than 10% accuracy. These measurements differ from those of Bakewell (1966) in that two components have been measured (whereas Bakewell measured only one), and in the much increased streamwise separation.

We plan to obtain the third component of velocity by use of the continuity equation. That is, after the correlation curves have been interpolated by Fourier series, they can be differentiated, and the unmeasured correlations involving the third component obtained by solution of the differential equation deduced from the continuity condition.

We agree that the amount of data necessary to carry out this sort of decomposition is large. However, we feel (an unproven assertion) that the amount of effort is not greater than that required to obtain equivalent information of the same statistical accuracy by any other means. Many of the measurements obtained by conditioned sampling are not statistically stable, being based on relatively small statistical samples.

The measurements in the sublayer were begun under the sponsorship of the General Hydrodynamics Research Program of the David W. Taylor Naval Ship Research and Development Center, and were completed with the support of the Fluids Engineering Unit of the Applied Research Laboratory at the Pennsylvania State University, where they were carried out. The data Analysis is being carried out with the support of The NASA Ames Research Center (under Dr. G. T. Chapman), making use of the computational facilities of the NASA Langley Research Center, with the collaboration of Dr. T. Gatski.

REFERENCES

Bakewell, H. P. (1966) The viscous sublayer and adjacent wall region in turbulent pipe flow. Ph. D. Thesis. University Park: The Pennsylvania State University.

Bevilaqua, P. M. & Lykoudis, P. S. (1971) Mechanism of entrainment in turbulent wakes. AIAA Journal 9: 1657-1659.

Bradshaw, P. (1966) The effect of initial conditions on the development of a free shear layer. J. Fluid Mech. 26: 225-236.

Brillinger, D. R. & Rosenblatt, M. (1967a) Asymptotic theory of estimates of Kth order spectra. Spectral Analysis of Time Series. (ed. B. Harris). pp. 153-188. New York: Wiley.

Brillinger, D. R. & Rosenblatt, M. (1967b) Computation and interpretation of Kth order spectra. Spectral Analysis of Time Series. (ed. B. Harris). pp. 189-232. New York: Wiley.

Browand, F. K. & Troutt, T. R. (1980) A note on spanwise structure in the two-dimensional mixing layer. J. Fluid Mech. 917: 771-781.

Brown, G. & Roshko, A. (1971) Symposium on turbulent shear flows, p. 23.1. London: Imperial College.

Brown, G. & Roshko, A. (1974) On density effects and large structure in turbulent mixing layers. J. Fluid Mech. 64: 775-816.

Cantwell, B. J. (1981) Organized Motion in Turbulent Flow, in Ann. Rev. Fluid Mech. 13: 457-515. M. Van Dyke, J. V. Wehausen, J. L. Lumley, eds. Palo Alto: Annual Reviews, Inc.

Chevray, R. (1968) The turbulent wake of a body of revolution. J. Basic Engineering (Trans. ASME Series D) 90: 275-284.

Chevray, R. & Kovasznay, L. S. G. (1969) Turbulence measurements in the wake of a thin flat plate. AIAA J. 7: 1641-1643.

Eisely, L. (1979) Darwin and the Mysterious Mr. X: New Light on the Evolutionists. New York: E. P. Dutton.

Hofstadter, D. R. (1980) Goedel, Escher and Bach. New York: Vantage.

Hussain, A. K. M. F. (1981) Coherent structures and studies of perturbed and unperturbed jets, in Proceedings, An international conference on the role of coherent structures in modeling turbulence and mixing. Madrid: University Politechnica. (To appear in Lecture Notes in Physics).

Lessen, M. (1978) On the power laws for turbulent jets, wakes and shearing layers and their relationship to the principle of marginal stability. J. Fluid Mech. 88: 535-540.

Lumley, J. L. (1967) The structure of inhomogeneous turbulent flows. In Atmospheric Turbulence and Radio Wave Propagation, A. M. Yaglom and V. I. Tatarsky, eds., pp. 166-178. Moscow: NAUKA.

Lumley, J. L. (1970) Stochastic Tools in Turbulence. New York: Academic Press.

Malkus, W. V. R. (1956) Outline of a theory of turbulent shear flow. J. Fluid Mech. 1: 521-539.

Malkus, W. V. R. & Veronis, G. (1958) Finite amplitude cellular convection. J. Fluid Mech. 4: 225-260.

Monin, A. S. & Yaglom, A. M. (1975) Statistical Fluid Mechanics. (J. L. Lumley, ed.) Cambridge, MA: M. I. T. Press.

Payne, F. R. (1966) Large eddy structrure of the turbulent wake behind a circular cylinder. Ph. D. Thesis. University Park: The Pennsylvania State University.

Payne, F. R. & Lumley, J. L. (1967) Large eddy structure of the turbulent wake behind a circular cylinder. Physics of Fluids. 10: S194-S196.

Rice, S. O. (1944) Mathematical analysis of random noise. Bell System Tech. J. 23: 282-332.

Rosenblatt, M. (1966) Remarks on higher order spectra. Symposium on multivariate analysis. New York: Academic.

Rosenblatt, M. & Van Ness, X. (1965) Estimation of the bi-spectrum. Ann. Math. Stat. 36:1120-1136.

Roshko, A. (1981) The plane mixing layer; flow visualization results and three-dimensional effects. In Proceedings, An international conference on the role of coherent structures in modelling turbulence and mixing. Madrid: University Politecnica. (To appear in Lecture notes in Physics).

Taulbee, D. & Lumley, J. L. (1981) Prediction of the turbulent wake with a second order closure model. J. Fluid Mech. To be submitted.

Tennekes, H. & Lumley, J. L. (1972) A First Course in Turbulence. Cambridge, MA: M. I. T. Press.

Townsend, A. A. (1947) Measurements in the turbulent wake of a cylinder. Proc. Roy. Soc. A 190: 551-561.

This work supported in part by the U. S. NASA-Ames Research Center under Grant No. NSG-2382, and in part by the U. S. National Science Foundation, Engineering Division, under Grant No. 79-19817.

Sibley School of Mechanical
and Aerospace Engineering,
Cornell University
Ithaca, NY 14853

Index

V

W